Judite Pinto Tene

Perception and Acceptability of Flood Risk in Downtown Ricatla

Judite Pinto Tene

Perception and Acceptability of Flood Risk in Downtown Ricatla

How Society Understands and Reacts to Environmental Risks in a Changing World

ScienciaScripts

Imprint

Any brand names and product names mentioned in this book are subject to trademark, brand or patent protection and are trademarks or registered trademarks of their respective holders. The use of brand names, product names, common names, trade names, product descriptions etc. even without a particular marking in this work is in no way to be construed to mean that such names may be regarded as unrestricted in respect of trademark and brand protection legislation and could thus be used by anyone.

Cover image: www.ingimage.com

This book is a translation from the original published under ISBN 978-3-639-61109-0.

Publisher:
Sciencia Scripts
is a trademark of
Dodo Books Indian Ocean Ltd. and OmniScriptum S.R.L publishing group

120 High Road, East Finchley, London, N2 9ED, United Kingdom
Str. Armeneasca 28/1, office 1, Chisinau MD-2012, Republic of Moldova, Europe
Managing Directors: Ieva Konstantinova, Victoria Ursu
info@omniscriptum.com

Printed at: see last page
ISBN: 978-620-3-34104-1

Index

List of Symbols and Abbreviations

BR	Boletim da República
CENACARTA	Centro Nacional de Cartografia e Teledetecção
EDM	Electricidade de Moçambique
EN1	Estrada Nacional 1
FCTA	Faculdade de Ciências da Terra e Ambiente
FIPAG	Fundo de Investimento e Património do Abastecimento de Água
GRA	Gestão de Riscos Ambientais
INE	Instituto Nacional de Estatística
IIAM	Instituto de Investigação Agrária de Moçambique
Km^2	Quilómetros quadrados
LT	Lei de Terra
MICOA	Ministério para a Coordenação de Acção Ambiental
OT	Ordenamento do Território
PSSA	Pequeno Sistema de Abastecimento de Água
RLOT	Regulamento da Lei de Ordenamento Territorial
RGPH	Recenseamento Geral da População e Habitação
RSU	Regulamento do Solo Urbano
SDPI	Serviços Distrital de Planeamento e Infra-estrutura
UPM	Universidade Pedagógica de Maputo

Dedication

I dedicate this work to my parents, Pinto Tene and Carlota Boné Bata;

To my siblings, José, Marcos and Carla Pinto Tene and their beautiful family;

To my daughters, Keren and Shenízia.

Thanks

To God, for his infinite and tireless blessings and for guarding and strengthening me at all times.

To my supervisor, Prof. Dr. Gustavo Sobrinho Dgedge, for his guidance, support, patience and teachings in the conception of this dissertation.

The same thanks go to the Faculty of Earth Sciences, particularly the Department of Environment and Development, for its patience with the first group of master's students in Environmental Risk Management.

To the IBE, for financing the Master's scholarship, I recognize that without your help this dream would not have come true.

To the community of Ricatla for their support during the data collection process and to the Marracuene District Administration. My gratitude also goes to all my classmates, who were there for me and provided all the support I needed during my journey towards my master's degree.

I would like to thank my godmother, Dr. Elisa Mandlate, the woman who has always fought for equal opportunities for women in higher education, for her great support.

To my daughters, Keren Basílio Assane and Shenízia Basílio Alberto Assane, my unconditional support, the ones who were always up all night waiting for their mother to come home and when difficulties arose, there was always that word of hope..." Courage mom, it will work out ".

My thanks go to Ramalho Machacule, my source of inspiration, the pillar and the beam that never lets me down, for having been part of this journey and for being that person who, even though he is far away, still takes the time to find out about me and give me strength whenever I need it.

To my great friend António Paulo for his support, strength and encouragement during the development of this research.

To my family, friends and colleagues, for allowing us to learn together about the different topics covered during the course, and who directly or indirectly contributed to this dream becoming a reality and were able to share all of life's good and bad moments with me.

Thank you so much!

Summary

The theme of this study is: Perception and Acceptability of Flood Risk in Baixa do Ricatla, Marracuene District. The main objective of this research was to analyze the perception and acceptance of flood risk by residents, exploring the relationship between socio-economic conditions, housing conditions and environmental factors affecting the community. The methodology employed was mixed, combining qualitative and quantitative approaches. Interviews were conducted with residents and data was collected through questionnaires to community and municipal leaders. The research included a literature review, direct observation and analysis of socio-demographic data and housing conditions. The results indicate that the population of Baixa de Ricatla is mostly young, male, with a low level of education and a marital or single marital status. Housing conditions are precarious and unequal, with a predominance of buildings made of wood, zinc and reeds. Many homes are incomplete and vulnerable to flooding, thus compromising the quality of life for residents. In addition, access to basic services such as drinking water and health care is limited. The perception of the risk of flooding among residents varies considerably. Around 82% of residents reported having already faced flooding, and 44% rated the risk as high. Acceptance of the risk is notable, with some residents treating flooding as an inevitable part of their lives and even as an economic opportunity for activities such as fishing. The study's recommendations include resettling families in low-risk areas, developing a plan to make these areas attractive and transforming unoccupied areas into spaces of public utility. In addition, the study suggests promoting reforestation projects and using durable construction materials adapted to local conditions to increase the resilience of buildings

Keywords: Perception, Acceptability, Risk, Flooding.

0. INTRODUCTION

This dissertation, on the theme of Perception and Acceptability of Flood Risk in Lower Ricatla, aims to understand the factors that contribute to the acceptability of flood risk in Lower Ricatla in the district of Marracuene, where people suffer every year and lose their homes and other property to floods.

In Mozambique, several studies deal with the risks of flooding that occur in low-lying areas occupied for the construction of houses and other high-value infrastructures. For the authors, perception is a crucial element in the study of risk and other phenomena that plague the population occupying low-lying and flooded areas throughout most of the country.

Nowadays, the perception and acceptance of risk is necessary for the occupation of any space, although population growth and economic and social conditions exert pressure when it comes to deciding where to live. This means that the individuals who occupy low-lying areas perceive the phenomenon according to their individual expectations and perspectives, reflecting in nature the experiences and aspirations of those who perceive and accept the phenomenon, risking any kind of damage that may occur.

Contrary to the widespread belief in various literatures in the country, most of these disasters do not occur naturally. Various environmental imbalances are evident, affecting the elements of the landscape (SILVA & POLERO, 2015). Human actions are intrinsically linked to factors such as social inequality, poverty, unemployment, wars and vulnerable health and education institutions, which are unable to adequately prepare humans to exploit the land in a sustainable manner and protect themselves from dangerous actions in relation to nature.

Recently, we have seen an increase in human vulnerability and in the intensity of the impact of environmental risks in our country. The advance of environmental degradation is closely linked to the low performance of public power and the difficulty of access to land and housing in urban areas, resulting in the intensification of occupation of unsuitable areas, such as steep slopes, floodplains and places prone to flooding (SANTOS, 2007).

Certain areas of the country are identified as particularly vulnerable, mainly due to the high water table associated with the inadequate accumulation of garbage, obstructing the passage of water. In the event of prolonged rainfall, these areas become prone to flooding. The recognition of this risk by the Disaster Management Institute (INGD) has led to these areas being marked; however, some people ignore these warnings, persisting in disorderly construction, as is the case in the Ricatla lowlands, in Marracuene District, Maputo Province.

It is crucial to consider that the population's insistence on settling in flood-prone areas leads to additional problems such as sanitation, particularly the invasion of homes by contaminated water resulting in the outbreak of endemic diseases due to cohabitation with rodents, cockroaches, mosquitoes, snakes and other animals, posing serious health risks.

In this context, this study aims to analyze the perception of risks and the factors behind their acceptability in downtown Ricatla. It seeks to understand the perception and behavior of individuals or social groups in the face of risks, addressing the acceptance and rejection of the risk of flooding, as well as measures of adaptation and coexistence with the constant presence of these risks.

The acceptability of risk emerges as a crucial factor, influencing and reinforcing the behavior of people living in flood risk areas, promoting the adoption of precautions. The experience of flood risk amplifies the perception of suffering that a family or individual may face during the rainy season (GOTHAM, 2018).

The study of risk is currently linked to perception, as this determines the way we act and influences the application of new technologies. The recognition of something as dangerous is influenced by the socio-historical context and individual perception present in the lives of all human beings, shaping their behavior and attitudes in their daily activities.

Faced with this reality, it is imperative that researchers and scholars of natural hazards intensify their research into disaster prevention, since the loss of human life and material goods occurs constantly. The measures adopted must, as a priority, be preventive, not just corrective (Cristo, 2004).

Encouraging environmental awareness through education enables citizens to become more aware of their actions, realizing the need to preserve natural resources for present and future generations. For environmental awareness to be effective, it is essential to identify the appropriate perception of the local reality, promoting the rational and sustainable use of natural resources (SOUZA, 2018: 45).

This dissertation is structured to include the introduction, which describes the problematization, the general and specific objectives of the research and the justification for choosing the topic. The first chapter presents the conceptual framework on related themes such as risks, vulnerabilities, acceptability of flood risk, perception of flood risk, as well as the informal occupation of space. The second chapter describes the methodology used. The third chapter describes the area and object of study as a basis for the dissertation, including geographical location, physical-natural and socio-economic characteristics. The fourth and final chapter consists of an analysis of the results and discussion, directing new paths and reflections for future research on the subject in question.

0.1 Problematization

In the last two decades, Mozambique has experienced significant growth in the construction of housing and infrastructure, especially in the flood plains. The National Institute for Disaster Management (INGD, 2014) highlighted that floods increased state spending by around 32%, contributing an additional 2 to 3 percentage points to the poverty rate, revealing the alarming vulnerability of affected communities.

Floods are recurrent natural events that have an acute impact on the lives of communities, especially those located in risk areas. The social acceptability of flood risk is shaped by how communities perceive and react to it, influenced by socio-economic and cultural factors and the effectiveness of risk management policies. The lack of a clear understanding of how different social groups perceive and accept these risks can result in inadequate policies and the perpetuation of vulnerability.

The normalization of the social acceptability of flood risk is evident, as many residents choose to build in risk areas, aware of the potential consequences. This acceptance is often associated with socio-economic interests, such as proximity to workplaces and markets, which lead individuals to ignore the annual risks of flooding.

In the Marracuene District, particularly in the lower Ricatla, the acceptability of the risk of flooding has become commonplace. People occupy these low-lying areas, aware of the future risks, oblivious to the annual damage. The area has witnessed a significant increase in population, despite the fact that floods result in serious damage, including material losses, human lives and agricultural crops. Reports from the head of the neighborhood, presented on Miramar news on March 21, 2024, revealed cases of drowning and loss of human life, this being the fourth such incident.

Surprisingly, despite the obvious risk, occupation continues to grow, justified by some residents due to the low cost of land, even though they recognize the area as prone to flooding. In addition, the population occupies dangerous spaces due to a lack of awareness of the existing threat or the perception that the event is unlikely to occur.

According to Televisão de Moçambique (TVM), approximately 700 families are at risk of flooding during the rainy season in the neighborhood of Ricatla, Marracuene, Maputo. The Director General of INGD, Augusta Maita, urged the local authorities to act quickly to remove this informal settlement (Jornal TVM, 23.06.2019), showing that this problem has persisted for a long time.

Despite the fact that the prohibition signs are known by the residents and recognized by the political-administrative structures of the neighborhood, there has been an increase in the number of constructions in the lower area, suggesting fraudulent occupation by invasion, transfer of homes between residents of the same neighborhood or illegal sale of land, in violation of article 3 of Law 19/97, of October 1 (Land Law).

Faced with these situations, the following question arises:

❖ ***What factors influence the acceptability of risk in Baixa do Ricatla in Marracuene District?***

0.2 Hypothesis

The acceptance, perception and permanence or return to the risk zones in downtown Ricatla are associated with the socio-economic situation and customs of the neighborhood's residents.

0.3 Justification

This research is an integral part of the Master's program in Environmental Risk Management (GRA), offered by the Faculty of Earth and Environmental Sciences (FCTA) of the Pedagogical University of Maputo. The scope of the study deals with the theme of "Perception and Acceptability of Flood Risk in Baixa de Ricatla, Marracuene District", with the aim of understanding the reasons that lead the population to accept living with the risk of flooding, in order to contribute to improving the local quality of life.

Increased urbanization and population growth have driven territorial expansion, leading to the occupation of areas that are often unsuitable for construction. This phenomenon requires in-depth analysis in order to understand and prevent environmental risks, seeking to ensure social security and the well-being of society in general.

The environmental, social and economic relevance of this research is justified by the need to identify the factors that influence the acceptability of risk and to propose measures to discourage construction in areas cyclically affected by natural disasters, the impact of which is increasing every year. Understanding the factors of acceptability will also contribute to understanding the underlying motives for occupation, even in the face of known challenges.

In addition, raising public awareness of the dangers of building in low-lying and vulnerable areas plays a crucial role in preventing water-borne diseases, especially during rainy periods, as well as other diseases transmitted by insects, rodents and snakes.

In the academic context, this study will provide a valuable source for literary consultations and will serve as a basis for future research on aspects related to the acceptability of urban flood risk, enriching the existing bibliographic repertoire on this subject. By addressing the occupation of areas unsuitable for construction, the study is aligned with the principles of sustainable development, promoting practices that aim to balance urban growth with environmental preservation and the safety of the population.

The results of this study can help in the formulation of effective public policies related to the occupation of risk areas. Understanding the factors of risk acceptability will provide valuable input for the implementation of preventive and awareness-raising measures.

0.4 Objectives

This dissertation aims to achieve the following objectives:

0.4.1. General objective

> Understanding the acceptability of risk in Lower Ricatla, Marracuene District.

0.4.2 Specific objectives

a) Describe the physical-natural and socio-economic aspects of the study area

b) Characterize the risk of flooding in the study area

c) Describe the perception of risk in downtown Ricatla;

d) Explain the perception and acceptability of risk in Ricatla.

CHAPTER 1: THEORETICAL AND CONCEPTUAL FRAMEWORK ON RISK FLOODS AND ACCEPTABILITY

This chapter presents the theoretical framework on the main concepts relating to flood risk analysis. In this way, the fundamental concepts for the research are delimited, including the concept of risks, floods, acceptability, vulnerability, exposure, hazard, as well as the factors associated with these elements.

1.1 Floods and flooding

According to FELIZARDO (2016), floods are natural phenomena that occur frequently in watercourses, usually triggered by heavy and rapid rainfall or long-duration rainfall. These natural events have been intensified, especially in urban areas, by anthropogenic activities.

According to WOLLMANN (2015), floods can be considered one of the consequences of the action and dynamics of natural systems on the earth's surface that cause the greatest changes in geographical space.

Flooding is the temporary increase in the level of water in the main course of the river, which then flows into the fullness of its smaller bed, but without overflowing onto the extensive floodplain (PDUL, 2020).

Floods are also conceptualized as natural phenomena, which occur periodically in watercourses due to high magnitude rainfall. These occur in urban areas and can be the result of intense rainfall with a long return period or due to the overflow of watercourses caused by changes in the balance of the hydrological cycle in regions upstream of urban areas; or even due to urbanization itself (BRAGA, 2016).

Flooding represents the overflow of water from a watercourse, reaching the floodplain or floodplain area, while floods are defined by the rise in water level in the drainage channel due to increased flow, reaching the maximum level of the channel, but without overflowing (IPT, 200 cited by TOMINAGA, et al., 2009).

According to CASTRO (2003), flooding can be defined as the overflow of water from rivers, lakes and dams. Floods are caused by the abnormal precipitation of water which, when it overflows from riverbeds, lakes, canals and dammed areas, invades adjacent land, causing damage. In densely populated areas, they can damage or destroy poorly located and unsound housing, as well as damaging furniture and other household items.

According to AMARAL & MONI (2020), flooding is the process that occurs when river waters overflow due to rainfall and occupy the area next to the river, which are called river plains or floodplains.

According to BRAGA (2016), flooding is a phenomenon that occurs when the waters of rivers, streams, rainwater galleries leave the drainage bed due to the lack of transport capacity of one of these systems and occupy areas where the population uses for housing, transportation (streets, roads and sidewalks), recreation, commerce, industry, among others.

According to UNISDR (2017), floods are the natural hazard with the highest frequency and the widest geographical distribution worldwide. Although most floods are small events, monstrous floods are not uncommon. Floods most commonly occur as a result of heavy rainfall when natural watercourses are unable to carry away the excess water. It can also result from other phenomena, particularly in coastal areas, from a storm tide associated with a tropical cyclone, a tsunami or a high tide.

Flooding is also understood as the rise in the level of rivers beyond their normal flow, causing their waters to overflow into areas close to them. These areas close to the river where the water overflows are called flood plains. When there is no overflow, even though the river is practically full, it is a flood and not a flood. For this reason, in the scientific world, the terms "flood" and "inundation" should be used differently (KOBIYAMA et al., 2006: 45).

BRAGA (2016) characterizes floods as a natural phenomenon that occurs when the flow to be drained is greater than the discharge capacity of the water system.

Flooding in areas occupied by human activities that are incompatible with the presence of water becomes a disaster with major socio-economic losses.

Flooding is caused by water that accumulates along streets due to heavy rainfall in cities with poor drainage systems. In floods, the flow of water depends much more on poor drainage, which hinders the flow of accumulated water, than on local rainfall (CASTRO, 2003: 51).

According to analyses by GRILO (1992) cited by BRAGA (2016), flooding generally occurs in flat areas or with depressions and valley bottoms, with surface runoff compromised by topography and the lack or insufficiency of a rainwater system in the urban environment. Furthermore, the smaller the expanse of green areas, the less water infiltrates into the soil, which feeds the suspended aquifers, causing less aid for surface runoff, which could mitigate the causes of flooding (TEODORO et al., 2007).

The distinction between the concepts involved in this study lies initially in understanding the factors that involve the dimension between flooding and inundation. There is therefore a common cause, the increase in flow, but as a consequence, each event has a certain result that is linked to specific conditions that involve other factors, which makes discerning these concepts more complex and often generalized.

Figure1 : Differences between flooding, inundation and waterlogging

Source: PDUL (2020).

The occurrence of floods results from the interaction of various factors that affect the formation and propagation of flows along the contributing watershed (CAMPOS, 2015: 69). These factors can be categorized as transient, associated with climatic events such as rainfall, evapotranspiration and soil saturation; permanent, related to the morphometric characteristics of the basin and geology; and mixed, linked to land use and occupation (COOKE & DOORNKAMP, apud CAMPOS, 2015).

According to TUCCI (1999), floods can occur in two ways:

a. In riverside areas: Rivers usually have two beds - the smaller one, where the water flows most of the time, and the larger one, which is flooded every two years on average. The negative impact occurs when the population occupies the larger riverbed and is subject to flooding. This type of flooding is a natural process, and the damage to the population is aggravated by occupying the flood valley during periods of drought or sequences of dry years.

b. Due to Urbanization: Urbanization increases the frequency and magnitude of floods due to land occupation with impermeable surfaces and drainage systems. Urban development can create obstacles to runoff, such as embankments and bridges, as well as contributing to inadequate drainage and obstructions to flow near pipes and siltation. This type of flooding is mainly the result of the inadequate design of drainage systems in cities and the sealing of surfaces, which increases surface runoff to the detriment of underground runoff.

According to TUCCI (2005), flooding occurs when heavy rainfall exceeds the soil's infiltration capacity. In this scenario, a significant part of the volume of water flows into the drainage system, exceeding the capacity of the smaller bed. This phenomenon is part of the natural hydrological cycle and is influenced by short-, medium- and long-term climate variability. The occurrence of these rainy events is random and depends on local and regional climatic processes.

Flash floods, on the other hand, are triggered by high-intensity rainfall concentrated in areas of rugged terrain or in urban environments. They are characterized by a rapid rise in water levels. This phenomenon is the result of the interaction of various atmospheric and terrestrial processes, including intense rainfall, soil moisture content, morphological characteristics of slopes, steep relief and the presence of impermeable surfaces. These elements combine to maximize water distribution capacity, contributing to the rapid rise in the level of water bodies (MACHADO, 2010).

1.1.1 Types and causes of flooding

According to BARBOSA (2006), the process of flooding occurs when the waters of rivers, streams and rainwater galleries leave the flow bed due to the lack of transport capacity of one of these systems and occupy areas that the population uses for housing, transportation, recreation, commerce, industry and others. These events can occur due to the natural behaviour of rivers or amplified by the effect of man-made changes in urbanization through the sealing of surfaces and the canalization of rivers.

Meteorological and hydrological conditions can increase the occurrence of floods. When rainfall is intense, the amount of water that arrives simultaneously in the river is greater than its drainage capacity, i.e. its normal channel, and therefore results in flooding. The problems resulting from flooding depend on the frequency with which they occur and the degree of occupation of the watershed by the population (TUCCI & BERTONI 2003).

Floods can be due to various causes and, depending on these, can be divided into several types, namely: river floods or floods, floods from topographic depressions, coastal floods and urban floods (RAMOS, 2013).

Table1 : Types of floods and their causes

TYPE	CAUSE
Flood (river flooding)	➢ Heavy and/or abundant rainfall ➢ Melting snow or ice ➢ Combined effect of rain + tidal effect and/or + *storm surge* ➢ Obstacles to river flow or collapse of obstacles
Flooding of topographic depressions	➢ Rise of the water table (natural or artificial) ➢ Retention of precipitation water by a soil or substrate ➢ Floods
Coastal flooding	➢ *Storm arises* ➢ *Tsunami* or *tsunami* ➢ Eustatic sea level rise ➢ Earthquakes with tectonic subsidence phenomena
Urban flooding	➢ Heavy rain, overloading of artificial drainage systems ➢ Rise of the water table (natural or artificial) ➢ Floods

Source: RAMOS, 2013.

1.1.2 Flooding processes in the urban environment

Flooding is associated with urbanization. SALINAS & ESPINOSA (2004) point out that since the dawn of humanity, human settlements have been located close to watercourses. Associated with this, deforestation and erosion have occurred and are still occurring, which together modify the hydrological response of the basins, thus increasing the occurrence and magnitude of floods.

According to BARBOSA (2006), as the population seals the soil and accelerates the flow through conduits and channels, the amount of water that reaches the drainage system at the same time increases, producing more frequent floods than those that existed when the surface was permeable and the flow occurred through natural gullies. The impacts on the urban environment resulting from flooding depend on the degree to which the floodplain is occupied by the population (riverine flooding) and the sealing and channelization of the drainage network (urban drainage).

SILVA & POLERO (2018) argue that urban flooding is normally caused by the dynamics of nature and is intensified by anthropic intervention in the environment. The socio-environmental effects are aggravated to the extent that the process of use and occupation of urban land is carried out in an inadequate manner, where the low-income population that occupies places unsuitable for housing is exposed to environmental and pathological risks generally present in risky housing sites" (SOUZA & ROMUALDO, 2020: 02).

Urban areas can suffer from river flooding, coastal flooding or flooding caused by rainwater and groundwater, depending on where the city is consolidated (BORGES, 2013). In general, the author considers that they originate from the combination of extreme meteorological and hydrological factors, without neglecting the anthropogenic action associated with disordered growth and occupation of accidental floodplains such as breaches and dams or reservoirs.

According to HOLZ (2010), urban flooding is a consequence of changes in land use and occupation that modify the urban hydrological cycle, since the percentage of water that infiltrates is reduced due to sealing, reduced surface roughness and the replacement of natural drainage channels with underground pipes or their grinding and lining.

According to MOTA (2003), changes in hydrological cycles are manifested by:

a) Increased precipitation, as human activities in cities produce a greater number of condensation nuclei;
b) Decreased evapotranspiration as a result of vegetation removal;
c) Decreased water infiltration due to sealing and soil compaction;
d) Increase in the amount of water from surface runoff;
e) Consumption of surface and underground water for public supply, industrial and other uses;

f) Changes in the water table, which may be reduced or depleted;

g) Increased soil erosion and consequent increased siltation of surface water collections;

h) Increased flooding and pollution of surface and groundwater.

1.1.3 Factors in the Occurrence of Flooding

The risk of flooding, as highlighted by CARVALHO & SARAIVA (2009), is the result of environmental dynamics, including changes in the physical conditions of the land, such as destruction of vegetation, compaction, sealing and changes in surface and groundwater drainage patterns, as well as the load exerted on lithological (geological) units. TOMINAGA (2009), cited by ALVES (2019), argues that the occurrence of floods is linked to the overflow of water, reaching flat areas adjacent to rivers, driven by factors such as the natural behaviour of rivers, urbanization, sealing of surfaces and canalization of rivers.

BORGES (2013) identifies key factors for the occurrence of floods, such as slow surface runoff in valleys and plains, in addition to anthropic actions in urban centers, which go beyond the occupation of prone areas. According to AMARAL & RIBEIRO (2009), cited by VESTENA *et al.*, (2014), natural disasters related to flooding result from the combination of natural factors, such as the shape of the relief, characteristics of the drainage network of the watershed, intensity, quantity, distribution and frequency of rainfall, soil characteristics, moisture content and the presence or absence of vegetation cover.

Anthropogenic factors include the irregular use and occupation of plains and the banks of watercourses, inadequate waste disposal near watercourses, changes in the characteristics of the watershed and watercourses (flow, straightening and channelization of watercourses, soil sealing, among others), as well as an intense process of soil erosion and silting up of watercourses.

Meanwhile, MONTE *et al.*, (2016), argues that the occurrence of floods is linked to the following factors:

a) **Geomorphological (Geomorphology)**: movements of slopes, such as landslides and mudslides that can reach the valley floor, blocking the river channel and causing flooding upstream (in the case of a river);

b) **Hydrological (logia)**: due to the rise of the water table at the bottom of a valley or topographic depression;

c) **Climatological (logia)**: through intense rainfall that can affect restricted areas, lasting a few minutes or hours and sometimes affecting the entire territory, lasting several days;

d) **Human (Anthropological)**: due to the construction of dams, leading to flooding upstream or the overtopping of dams leading to a flood collapse. It can also be said that the construction of sealing walls influences the occurrence of floods;

e) **Pedological (logia)**: when the soil loses its infiltration capacity due to trampling;

f) **Marine**: the effects of transgression or increased waves in marine waters that can mainly affect areas of low sea level.

1.1.4 Impacts caused by flooding

The impacts caused by floods can be classified as tangible and intangible, and at a second level, direct and indirect. Tangible assets can be measured in monetary terms. This does not apply to intangible assets, since these result from the physical contact of floodwater with goods and people, and are considered direct. In any case, both have implications for social and economic activities during and after floods (KOBIANA, MENDONÇA & MORENO, 2006).

Table2 : Classification of flood damage

Type of damage	Direct damage	Indirect Damages
Tangible Damage	• Physical damage to homes: construction and contents. • Physical damage to shops and services; building and contents (furniture, stock, goods on display, etc.). • Physical damage to equipment and industrial plants. • Physical damage to infrastructure.	• Cleaning costs, accommodation and medicines; • Reallocation of time and expenses in reconstruction; • Loss of income; • Loss of profit, loss of database information; • Additional costs for companies to create new operational routines. Multiplier effects of damage in interconnected economic sectors; • Interruption of production, loss of production; • Disruptions, stoppages, congestion in services, additional transport costs.
Intangible Damages	• Injury and loss of human life. • Diseases caused by contact with water, such as colds and infections. • Loss of objects of sentimental value; loss of animals.	• Psychological states of stress and anxiety; • Long-term damage to health; • Lack of motivation to work; inconvenience of interruption; • Disruptions to economic activities, means of transport and communication; • Disruption to residents' daily lives.

Source: Matlombe (2019).

Various studies on floods, such as those carried out by Khalequzzaman (1994), Ema (2002), El Raey & Beoro (2003), Jinchi (2005) and Asante et al. (2007c), categorize the impacts of this phenomenon into two distinct phases. The first phase refers to impacts during the flood event, characterized as negative due to their destructive nature. The second phase deals with the impacts that occur after the flood event. In this stage, not only negative factors are observed, but also positive ones, which can ultimately have productive effects for both the environment and society.

1.1.3.1 Negative impacts

According to Khalequzzaman (1994) and Asante *et al.*, (2005) floods are one of the most widespread and destructive natural disasters and, although many of the effects can be attributed to the natural events themselves, others are also associated with poor land use management. Flooding occurs naturally along most river systems and in low-lying areas.

Floods that occur in densely populated urban areas have the capacity to cause a great deal of damage to life and property (EWC *et al.*, 2003 and Chiesa *et al.*, 2012). The damage caused by floods consists of primary and secondary effects (Murwira, 2006).

Primary effects include: loss of human and animal life; physical damage to property and infrastructure (such as bridges, roads and sewage systems). Primary effects also include erosion of agricultural land and destruction of crops (El Raey and Beoro, 2003 and Jinchi, 2005).

Secondary effects occur after the floods have subsided and can include failures in water supply and sanitation infrastructure or contamination of water sources, both of which can lead to unhygienic conditions and the spread of disease (Murwira, 2006 and Moel *et al.*, 2009).

1.1.3.2 Positive impacts

Flooding can also have many positive impacts (Beilfuss and Dos Santos, 2001; El Raey & Beoro, 2003 and Jinchi, 2005). These include: recharging natural ecosystems; providing abundant fresh water for agriculture, health and sanitation; and depositing nutrient-rich sediments in various areas, thus increasing crop yields (Beilfuss and Dos Santos, 2001; El Raey and Beoro, 2003 and Jinchi, 2005).

According to Bonifacio (2022), although floods are generally associated with negative impacts, such as material damage and loss of life, there are some positive impacts that can arise in certain situations and contexts:

- Floods can deposit nutrient-rich sediment in the affected areas, enriching the soil and making it more fertile for cultivation.

- Flooding can help recharge underground aquifers, which are important sources of drinking water in many regions.

- **Promoting biodiversity:** Some species of plants and animals depend on seasonal floods for their life cycle. Periodic flooding can create or maintain important habitats for aquatic and terrestrial species.

- **Renewal of aquatic ecosystems:** Flooding can restore and revitalize aquatic ecosystems such as marshes and wetlands, which are vital for water filtration, flood mitigation and wildlife support.

- **Creation of recreational opportunities**: In some areas, flooding can create new opportunities for recreational activities, such as canoeing, fishing and bird watching, as new temporary bodies of water are formed.

- **Awareness of risk management:** After floods, there is often an increase in awareness of the importance of risk management and adaptation to climate change, which can lead to investments in more resilient infrastructure and more sustainable planning policies.

1.2 Flooding in Mozambique

According to MICOA (2005), in Mozambique, floods are caused by a number of factors, including intense localized rainfall, tropical cyclone activity, and poor dam management either in the national territory or in upstream countries. The extent of the areas affected by flooding depends on the rainfall pattern in Mozambique and its neighboring countries: Zimbabwe, Zambia, Malawi, South Africa and Swaziland.

According to Coamba (2018), the analysis of these floods in Mozambique takes into account the different altitudes of the country in relation to rising river water levels, providing estimates of the areas and populations potentially affected. In addition, poverty plays a crucial role in the perception of risk, significantly influencing people's decisions to occupy or abandon areas susceptible to dangerous or risky events. In urban areas, for example, risk spaces are often occupied by low-income populations, who rely on informal occupations for their livelihoods, without access to safe land on which to build homes.

Therefore, it is possible to say that the exposure of the Mozambican population to natural risks, such as floods, can be characterized as voluntary and forced. It is voluntary because communities perceive that the benefits of occupying these areas outweigh the associated costs and impacts, while it is forced because, for many of these communities, occupying these risk areas is the only viable alternative for accessing land and establishing permanent residence. This complex dynamic underscores the importance not only of disaster management, but also of socio-economic development and poverty reduction as essential components of community resilience in the face of the challenges of floods and other extreme weather events (Coamba, 2018: 77).

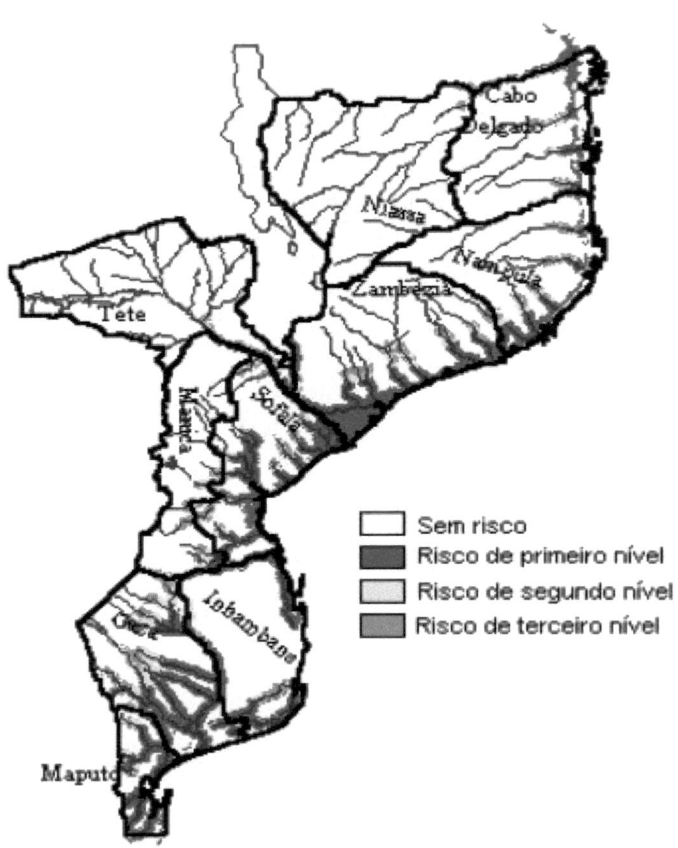

Source: INGC 2012

According to INGC (2012), river basin flooding is the most significant type of risk in the country. It is monitored by the National Water Directorate (DNA) on the basis of daily records of its flows up to critical limits, which can constitute alert levels. It is on the basis of this approach that the different degrees of flood risk can be determined based on altimetric classes, precipitation levels, watershed exposure and river flow levels.

a) **The first** degree **risk** affects 1.7 million hectares within an altitude of less than 20 meters above sea level, and within 10 km of the main river basins, representing around 6% of the national coverage. These areas can be flooded in a year of average to good rainfall.

b) **The second** degree **risk** is defined for an altitude of 20 - 50 meters above sea level, representing 2.7 million hectares (an estimated 9.6%) of the national dryland cover.

c) **The third** degree **risk** is defined for an altitude of 50 - 100 meters, with 10 km of approach to the main rivers and represents almost 4 million hectares (14% of the national land cover).

d) **The fourth** degree **risk** is less likely than the other three types, and will only have an effect in years when the flood risk is associated with water from regional river flows.

1.3 Flood risk management

Faced with environmental imbalances and humanitarian damage resulting from the   occurrence   of the occurrence of floods, a harmonious coexistence is possible based on initiatives   to   contain   and prevention of flood risk (Figueiredo, Velente, Coelho & Pinho, 2005).

According to Graciosa & Mendiondo (2018), flood risk management is a process that aims to reduce damage from floods by mitigating exposure, vulnerability and threat, acting in four integrated phases: before (1), during (2) and after (3) the event and mitigation (*See figure* 6).

Figure3 : Stages of flood risk management.

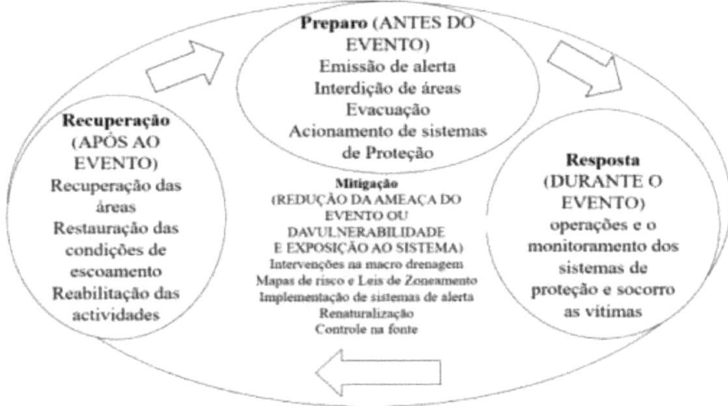

Source: Graciosa and Mendiondo (2018).

According to Monte, *et al.,* (2016) and Borges (2013), flood risk management involves

structural and non-structural measures. In turn, Borges (2013) states that for a good prevention strategy, these measures must act in a complementary way.

1.4 Flood mitigation and control and prevention

In order to make it possible for societies to live with floods, control measures are taken, which can be social, economic and administrative engineering actions. All of these aspects together result in a flood protection plan that aims to reduce the risks and the damage caused by floods, seeking sustainable urban development (Costa, 2013).

According to Tucci, Porto and Barros (1995), the principles for urban flood control are:

a) Consider the basin as a system (measures must not reduce the impact of one area to the detriment of another);

b) Analyze the control measures in the basin as a whole (structural and non-structural);

c) Consider the Urban Master Plan, Municipal, State and Federal Legislation and the Drainage Manual as the means of implementing flood control;

d) Do not extend the natural flood;

e) Follow the technical standards for the design and execution of drainage works.

Control measures can be classified as structural or non-structural and are generally applied in combination (Costa, 2013).

1.4.1 Structural measures

According to Costa (2013), structural control measures are engineering interventions that strategically modify the river system in order to reduce the risk and damage caused by floods.

For Barbosa (2006), structural measures are when man modifies the river through hydraulic works such as dams, dikes and canalization, among others. They can be classified into two types:

a) **Extensive measures** are those that are applied in the watershed, they aim to modify the relationship between precipitation and flow, an example is modifying the vegetation cover of the soil, in order to delay flood peaks and control erosion in the watershed (Barbosa, 2006).

b) **Intensive measures** are those that act directly on the water resource and which, according to Canholi (2005), can be classified into four types: accelerating runoff; slowing runoff; diverting runoff; and individual actions to make buildings flood-proof.

1.4.2 Non-structural measures

According to Barbosa (2006), non-structural measures, when man lives alongside the river, are preventive measures, such as zoning flood areas, a warning system linked to civil defense and insurance.

Measures cannot completely control flooding, but they can minimize its consequences. The combination of structural and non-structural measures ensures that the population suffers as little damage as possible, as well as enabling harmonious coexistence with the river. For riverside populations, this coexistence is essential to avoid material losses and even, in some cases, human losses. Decision-making is defined according to the characteristics of the river, the benefit of reducing flooding, as well as the social aspects of its impact (Barbosa, 2006).

1. 5. Risk

According to Gondim s/a: (87 - 88) quoted by Spink (2001:1279), the word risk appears around the 14th century in what is known as para-modernity and only in the 16th and 17th centuries does it emerge in the lexicons of the Latin and Anglo-Saxon languages respectively, with the modern meaning of representing "the possibility of future events occurring, at a time in history when the future was being thought of as subject to control".

Silva (2017) states that the word risk derives from the ancient Italian *risicare*, which means to dare. Initially, it was associated with the dangers of humanity, both natural disasters and fatalities and wars. However, in modern times, this measure of uncertainty is radicalized and takes on new contours, becoming society's essential reference in the fields of epidemiology, public health, technology, the environment, work and leisure-related activities" (Bernstein, 1997; Le Betone, 2012; Spink *et Al.,* 2004 cited by Silva 2017: 02).

Risk as a concept is understood in its strictest sense as hazard or aléas, which designates the spatial and temporal probability of a phenomenon occurring, in this case an undesired phenomenon, due to its negative consequences for man and society (Cunha and Dimuccio, 2002).

For Robert Charette (1989), the definition of risk shows that it refers, firstly, to future events; secondly, it involves change and, thirdly, it involves choice and the uncertainty that choice itself involves. For this reason, Aroça (2018) defines risk as any natural or man-made process that affects life and society's assets, while Borges (2013) understands risk as the probability that an event will occur that causes damage to either the population or the environment, together with the quantification of the serious consequences and economic costs, due to a natural or man-made phenomenon.

According to Cocurullo (2003: 45), risk is any situation that can affect the ability to achieve objectives. The term "risk" has been used as the possibility of a scenario occurring, whose effects are harmful, as a result of natural action and intensified by man.

Having analyzed the conclusions reached by these authors, we can point out that their ideas converge in the combination of the probability of occurrence of a certain undesirable phenomenon, for human health, the environment and economic activities associated with flooding.

Figure4 : Representation of the general technical concept of flood risk.

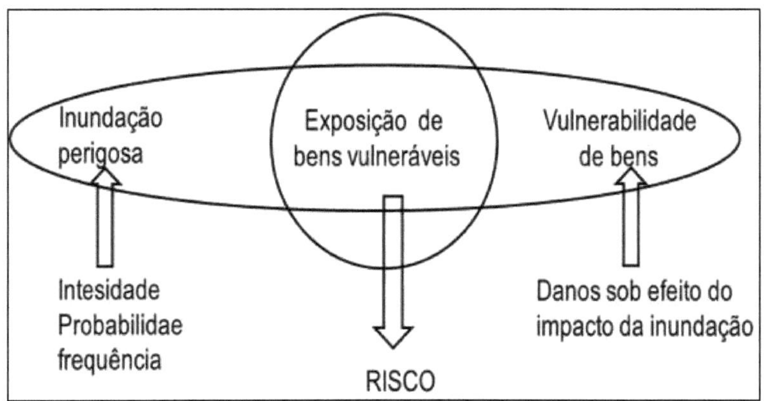

Source: Sousa (2012).

The typology of risks can be found in the following table:

Table3 : Types of risks

Risks			
Natural Risks	**Social Risks**	**Mixed Risks**	**Technological Risks**
"They result from the association between natural risks and risks arising from natural processes aggravated by human activity and the occupation of territory" (Richemond, 2007).	"They result from the combination of a large number of variables, which are particularly difficult to consider at the same time" (Dagnino and Júnior, 2007 citing Vieillard-Baron, 2007).	They occur when the phenomenon causing the damage has combined causes, or through the combination of risks of human origin.	"They correspond to sudden or unplanned accidents that result from or are induced by human activities" (Borges, 2013).

Source: Adapted by the author, 2024.

As can be seen from the table above, the concept of risk shows that it is the probability or uncertainty that results from a threat or danger that can arise in various ways, whether of natural or human-induced origin, which happens when people are exposed and vulnerable to the occurrence of a particular event.

Mathematically, the risk can be expressed as follows:

$$R = P * E * V$$

Where

P = probability of the hazard occurring;

V = value of the elements at risk, which is a function of the development in the exposed areas, the land use and the probability of presence (exposure); e,

E= the susceptibility of the elements at risk, which is a function of the extent of the hazard, as well as the socio-economic construction of the exposed elements (vulnerability).

Figure5 : Risk as a combination of Hazard, Vulnerability and Exposure

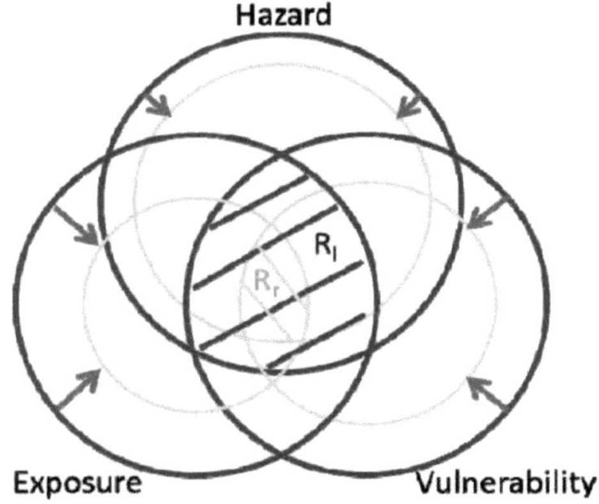

Source: WMO, 2013.

It is important to emphasize that there are only risks when there is an individual or a population that perceives and feels that they are threatened by this danger and could suffer the effects of this threat (Bento, 2012).

1.6. Risk acceptability

The concept of risk acceptance can be defined as the process of collecting, selecting and interpreting signals related to the uncertain impacts of events, activities and technologies (Wachinger and Renn, 2010). This process is complex, dynamic and influenced by various factors, such as knowledge, experience, values, attitudes and feelings, which interfere with people's thinking and judgment about the seriousness and acceptability of risks (Slovic, 2010; Wachinger and Renn, 2010).

According to Bezerra (2011), risk acceptability criteria are an important reference for assessing risk reduction measures and drawing up emergency response procedures, based on safety objectives and taking into account the particularities of each activity. Updating these criteria is important for the continuous improvement of safety conditions. The risk acceptability criteria are an important reference for Safety, Environment and Health (SHE) management. The results of the risk assessment will always be related to some level of uncertainty which may be very or somewhat relevant, depending on the model used for the estimates, the simplifications and the data assumed from a more or less sparse database.

The decisions to assume the appropriate risk acceptability criteria need to follow this order:

1) Be suitable for decision-making to reduce risk;

2) Adequate communication;

3) Formulating unambiguous criteria;

4) Be independent of concepts that favor the acceptability of risk.

Authors such as Domènech, Supranamiam and Sauri (2010) highlight the importance of risk information in increasing people's concern about adopting adaptive measures in their future behavior. However, Susan and Myers (2009) emphasize that the discussion about risks goes beyond information to include issues such as power, social control, ethics, inclusion and personal values.

1.6.1 Acceptable risk

According to Almeida s/a, the concept of socially acceptable risk refers to the conscious decision to take a sufficiently low risk, considering the probability of an incident or exposure related to the danger occurring, as well as the severity of the damage that could result. This level of risk is considered tolerable in a specific situation.

When people learn about their risks and understand the need to protect themselves, this opens the way for collaboration in prevention and protection actions.

It is understood that the social acceptance of risk involves the deliberate choice of individuals to assume a specific risk, recognizing that its effects can be managed without treatment or are subject to monitoring. By recognizing risk as acceptable, people admit that there is no zero probability of occurrence, and in the analysis of costs and benefits, they may choose to live in risky areas.

According to D'ARAÚJO (2013), socially acceptable risk constitutes the possible losses that could be accepted by a community in exchange for a certain profit or benefit, recognizing that it is not possible to completely eliminate all risks from hazards.

This perspective implies a pragmatic understanding that, in certain situations, it is impractical or economically unfeasible to completely eliminate risk. Therefore, by accepting a certain degree of risk, individuals and communities are aware of the measures needed to mitigate or deal with possible adverse impacts, thus promoting a more balanced approach to risk management.

Figure6 : Acceptable and unacceptable risk

Source: MANUELE. F. (2010).

1.7 Perception

According to ALÍPIO & VALE (S/d), perception is a complex phenomenon through which the outside world is apprehended and interpreted as being ordered into totalities. For the authors, previous experiences and stimuli are what characterize perception

Perception is defined as the process of organizing and interpreting data, situations, processes that result from the subject's own awareness and the environment with a representation of external objects or facts (Matos & Jardilino, 2016).

Perception is thus a selective and exploratory alternative that consists of understanding the environment based on sensations, with some aspects being registered and others ignored. The ability to understand the environment is related to cognitive activities through sensory stimuli, i.e. sight, hearing, touch, taste and smell (Klais, 2012).

According to Messner and Meyer (2005), the perception of a risk varies from one individual to another and from one social group to another, taking into account the level of information, uncertainties, intuitive behavior, and specific power constellations, as well as positions of interest.

Risk perception is precisely a person's ability to recognize situations or conditions of exposure to risk in the workplace, as well as to identify the frequency with which they are exposed to the danger determined by the risk and to quantify the intensity of this exposure. In order to have a good perception of risk, it is important that people undergo specific training on risks related to their routine and living with floods, observing the tolerance limits for the risks of accidents to which they are exposed and their effective form of prevention (Fernando, 2022).

Dook and Lognecker (2004) state that risk perception is determined by a number of factors, including internal motivations, previous experiences, assumptions about environmental conditions and the rate of change of a situation.

Meanwhile, Araújo and Santos (2020) consider that perception consists of the action of seeing and interpreting the behavior of what is observed, given to the individual or phenomenon analyzed.

1.7.1 Risk perception

Abreu and Zanella (2015) point out that risk perception is the ability to interpret a situation of potential harm, based on previous experiences and their extrapolation to a future moment, an ability that varies from a vague opinion to a firm conviction.

According to Souza (2015), many decisions about risk and other things in life are influenced by perception. According to the authors, an individual's perception of risk is defined by factors that deem the risk acceptable or to be avoided. These include the degree to which the risk is known or unknown, threatening or attractive, voluntary or involuntary, controllable or uncontrollable. This view is influenced by our experience, expectations, needs and the media and information to which it is exposed. In some ways, decisions are based on a combination of subjective reasoning and objective experience to determine risk; the least we can or should do is know where we've come from in the past, measure how well we're doing in the present and predict what might happen next, or in the future.

The notion of risk perception varies from one individual to another, taking into account the level of knowledge, uncertainties, interests and behaviors. In this way, it can be understood that when faced with risks, people select those to which they should attach importance and those to which they should not. Thus, understanding the link between social and risk perception helps us to understand how an individual or group perceives the occurrence, or not, of a particular risk related to natural and human adversities (Abreu & Zanella, 2015).

Messner and Meyer (2005) state that the perception of risk varies from person to person, with some individuals having lower levels of perception and tending to experience the risk and caring little about adopting prevention mechanisms because they believe that the risk may occur [...]. The same authors emphasize that individuals with lower levels of perception tend to experience the risk and do not adopt preventive mechanisms, as they do not believe that the risk can occur and at the same time are more vulnerable to floods.

Injage (2013) analyzes the perception of flood risk based on three types of perception:

- Probabilistic perception: accepts that disasters will occur and also perceives a pattern of occurrence;

- Dissonant perception: denies the threat of danger or previous events;

- Determining perception: recognizes the existence of catastrophes, but tries to see extreme events as occurring cyclically and irregularly.

1.7.2 Perception of Flood Risk

De Kellens et al., (2011), highlights a fundamental role in the success of the flood risk management process. He also states that it consists of assessing the probability of danger and the probability of results perceived by society.

In turn, Injage (2013) states that the perception of flood risk is linked to the way in which individuals analyze the environment, from the point of view of cultural evaluation, organization and characteristics of the lived space. Thus, individuals not only consider the probability of flooding occurring, but also the estimated severity and extent of its effects, and this may be the context through which the risk experienced is what determines its perception.

It should be noted that just by living in a low-lying area, the individual should already be aware of the likelihood of floods happening at some point during the rainy season (probabilistic perception), without forgetting that the history of flooding in downtown Ricatla has been reported in the country's news ever since that region became inhabited (dissonant perception) and finally we can consider that the majority of the population is aware of the tropical storms that affect the country cyclically and specifically the floods in downtown Ricatla, but people believe that these events are occasional and can occur anywhere else during the rainy season and not necessarily the region in which they live (determinant perception).

1.7.3 Factors affecting risk perception

The main factors that influence the perception of a risk include familiarity with the risk, education, experience, voluntary acceptance of the risk, ability to control the risk, catastrophic potential of the risk and the estimated forecast of its effects, impact on future generations, sensory perception, location, culture, use of space and the (ir)reversible effects of the risk (Giulio et al, 2015; Lechowska, 2018).

a) **Familiarity with the risk**: this refers to the ability to tolerate and live with the risk, and can be greater when the frequency and possibility of the risk occurring are also greater.

b) **Schooling**: people with schooling are more enlightened and readily admit the existence of risk than those with less schooling.

c) **Experiences**: risk perception is based on knowledge, which in turn is based on previously obtained information. Thus, ignorance of the causes of flooding diminishes perception; on the other hand, living in susceptible areas or forced relocations as a result of a risk can cause a sense of loss.

d) **Voluntary acceptance of risk**: the ability to accept and live with risk depends on how the gains or losses are perceived in the place of risk.

e) **Risk control capacity**: the existence of appropriate technical and institutional mechanisms that determine how to live with a given hazard. Thus, the greater the confidence in control and management instruments, policies and techniques, the greater the capacity for coexistence.

f) **The catastrophic potential of the risk and the estimated predictability of its effects**: the greater the likelihood of a catastrophe occurring, the lower the capacity for coexistence and tolerance in the face of this circumstance.

g) **Impact on future generations:** this is related to environmental sustainability and the awareness that present actions may have consequences for future generations and their opportunities.

h) **Sensory perception**: risks that are not immediately perceptible through sensory experiences tend to be rejected, rather than those whose effects are immediate and visible.

i) **Location**: proximity to the affected areas gives individuals a broad perception of the risk. Thus, perception tends to be greater if individuals live in areas close to the coast, low-lying areas and flood plains, where the effects of precipitation have been visible.

j) **Culture**: even if they live in areas considered to be at risk, individuals have a strong attachment to a certain place and understand that the environment where they were born and raised is fundamental to the continuity of their identities.

k) **Use of space**: the context in which a piece of land is used influences the perception of environmental risks (housing, farming, market).

l) **(Irreversible) effects of the risk**: irreversible risks are less capable of being dealt with than reversible ones.

Combining these factors, it can be considered that the population most exposed to the risk of risk of flooding are those who live on the downstream banks of rivers, conditioned by the lack of and search for better living conditions, fertile land and cities.

1.8. Vulnerability

Vulnerability, a central concept in risk and disaster management, is defined as the conditions determined by physical, social, economic and environmental factors that increase the susceptibility of communities to the impact of hazards or the occurrence of disasters, as established by Law 10/2020 of August 24 in Brazil. This broad and comprehensive definition highlights the interconnection of various elements that contribute to vulnerability.

Broesrsma et al. (2003) cited by MICOA (2006:12), broaden this understanding by referring to vulnerability as the degree of susceptibility or inability of a system to respond to the adverse effects of climate change, including climate variability and extreme events. This perspective considers the dynamics of climatic conditions as aggravating factors of vulnerability.

Pelling (2003) breaks vulnerability down into three main categories: physical (related to buildings), social (linked to the social, economic and political system) and human (a combination of physical and social factors). This multifaceted approach highlights that vulnerability is a complex condition that encompasses several interconnected dimensions. MICOA's (2007) perspective on vulnerability highlights the impact of climate change on the system, underlining both the sensitivity of the system and the capacity to adapt to new climatic conditions as crucial components of this vulnerability.

Bazzan (2011) contributes by defining vulnerability as part of a system that encompasses not only physical and moral damage to people, but also potential damage to exposed elements such as the production of socio-economic goods and heritage. This view emphasizes the breadth of the consequences associated with vulnerability.

The growing consensus in recent decades, mentioned by Carmo & Valencio (2014), IPCC (2012), UN-HS (2011), Welle & Birkman (2013), Wisner et al. (2013) and WMO (2015), highlights that the magnitude of a disaster is intrinsically linked to the vulnerability and exposure conditions of the population in risk areas. This highlights the crucial importance of integrated approaches to risk management.

Cardona (2003) points out that vulnerability expresses the degree of loss to which a given element is subject if the phenomenon in question occurs. This expression of vulnerability highlights the dimension of the consequences associated with adverse events. Therefore, by cross-referencing the perspectives of various authors, we realize that vulnerability is a complex and multifaceted concept, involving not only exposure to risk, but also the sensitivity and adaptability of social, economic and environmental systems in the face of adverse events. This integrated understanding is essential for developing effective risk reduction strategies and building resilience in vulnerable communities.

1.8.1 Types of vulnerability

This section will present the types of vulnerability, which are natural, environmental and socio-environmental vulnerability.

> **Natural vulnerability**

According to Klois *et al,* (2012) natural vulnerability shows the pre-disposition of the environment to natural factors. Examples: geomorphology, geology, soils and their stability in relation to morphogenesis and pedogenesis.

> **Social Vulnerability**

Cannon *et al.* (2003) propose that social vulnerability is a complex configuration of characteristics that include personal well-being, means of subsistence, resistance to adverse events, self-protection and political, social and institutional networks. This level of well-being of individuals, communities and society includes aspects related to the level of education, schooling, security and public policies, respect for human rights, social equality, among others (ISDR, 2004).

According to Hill and Cutter (2001), social vulnerability describes the demographic characteristics of different social groups that make them more or less susceptible to the negative impacts of an extreme event. For these authors, social vulnerability suggests that people create their own vulnerability through their actions and decisions.

> **Environmental vulnerability**

Klois et al. (2012) defines environmental vulnerability as any susceptibility of the environment to a potential impact caused by any anthropic use. Examples: disorderly construction and removal of vegetation cover.

> **Socio-environmental vulnerability**

Cartier Ruy (2009) conceptualizes socio-environmental vulnerability as a coexistence or spatial overlap between poor, discriminated against and highly deprived population groups (social vulnerability), who live or circulate in areas of risk or environmental degradation (environmental vulnerability), and in these regions the poor and discriminated against population is forced to live as "sacrifice zones".

Vulnerability and its relationship with poverty and lack of information have therefore been considered, since no one lives in an area of risk or environmental degradation on a whim, which means that there is a greater force driving the population to live in environmentally vulnerable areas.

1.8.2 Vulnerability indicators

According to Pine (2008), indicators play a fundamental role in offering a quantitative representation of phenomena, and are crucial for understanding a community's capacity to cope with, absorb or recover from disasters. Several academics, such as Birkmann (2006) and Dwyer et al. (2004), propose specific criteria for the selection and application of vulnerability indicators, emphasizing:

a) **Measurability:** Indicators must be capable of quantitative measurement.

b) **Relevance:** They must be directly related to the problem under study.

c) **Comprehensibility:** They must be easy to understand for different audiences.

d) **Easy interpretation:** Your interpretation should be accessible and clear.

e) **Analytical and Statistical:** They should allow for analytical and statistical studies.

f) **Data Availability:** The information needed for the indicators must be accessible.

g) **Comparability:** Indicators must be comparable between different geographical areas or time periods.

h) **Validity/Accuracy:** They must be valid and accurate.

i) **Reproducibility:** Must be reproducible in other studies.

j) **Conformity with the Research Problem:** They must be relevant to the specific problem in question

k) **Simplicity:** They must be designed in a simple and applicable way.

Coppola (2007) broadens the perspective, incorporating a variety of factors into the analysis of vulnerability, such as religion, race, gender, health, illiteracy rates, public policies, human rights, social inequalities, culture and tradition.

For their part, Cutter et al. (2000) propose a specific approach, using variables such as total population, total number of households, number of females, number of non-white people, number of people under 18, number of people over 65, average home value and number of mobile homes to measure vulnerability.

The diversity of criteria and variables highlights the complexity of vulnerability assessment, emphasizing the importance of an integrated approach that considers multiple factors for a complete and accurate analysis.

1.9 Exposure

As defined by the World Meteorological Organization (WMO, 2013), exposure refers to the presence of people, property, systems or other elements in risk zones, subject to potential loss. Measures of exposure can include the number of people or types of property in a given area.

According to Coamba's perspective (2018: 27), in Mozambique, the population is exposed to natural risks, in particular the risk of flooding, due to the lack of formal land use plans, as well as the lack of effective enforcement. This makes the marginal areas of urban centers, such as floodplains or steep slopes, prone to illegal and uncontrolled occupation, often considered abandoned land. The population's exposure to natural risks, such as the risk of flooding in Mozambique's urban centers, is a direct consequence of these informal and uncontrolled occupations of urban land.

In the area under study, it was observed that after signs were put up prohibiting the installation of various infrastructures, the competent authorities (Marracuene municipality) completely abandoned the site. However, this abandonment made it easier for the population to return and resettle in defiance of the previously established restrictions.

This dynamic illustrates the complexity of natural risk management in urban contexts, highlighting the need not only for formal policies, but also for effective implementation and continuous enforcement to mitigate the population's exposure to potential natural disasters. The cycle of informal and uncontrolled occupation highlights the urgency of more comprehensive and sustainable strategies to deal with growing exposure to natural threats.

1.10. Danger

According to Castro (2012), hazard is associated with the frequency of occurrence of a process and the place where it takes place. Hazardousness is defined as the occurrence of a process, with a certain level of intensity or severity, in a given time interval and in a specific location. It can be determined from the return period T (corresponding to the repetition of 2 events or processes that manifest themselves with the same characteristic).

Hazardousness is equivalent to what is known in Anglo-Saxon literature as *hazard*. According to the definition given by Varnes (1984), hazard is the probability of potentially destructive phenomena occurring in a given period of time and in a given area.

According to Verde (2008), dangerousness refers to the state or quality of being dangerous, representing the potential to cause harm, damage or risk. It encompasses two dimensions: time and space, i.e. dangerousness occurs in a given period of time in a given physical territory

1.11. Legislation associated with risk perception and acceptability

Consideration of risk perception and acceptability is fundamental when analyzing the possibilities for action both in reducing the probability of occurrence and the consequences in their various dimensions. A key principle of these guidelines is minimizing impacts on people, economic losses, environmental damage and maintaining the continuity of social functions in areas susceptible to flood risks.

Legislation associated with risk perception and acceptability comprises a set of laws and regulations that address how society deals with risks linked to different activities, products or processes. Risk perception is subjective and varies between individuals and communities. Risk acceptability relates to people's willingness to tolerate certain levels of risk in exchange for perceived benefits.

Documents covering matters relevant to legislation associated with risk perception and acceptability can include various areas, such as natural risk management, environmental safety, urban planning and emergency management. Thus, a brief list of them includes:

Water Law No. 16/91, approved by Parliament in 1991:

   a) Regulates the legal regime of inland waters;

   b) It defines the water resources that belong to the public domain;

   c) Determines the government's jurisdiction over water in the public domain;

   d) Establishes the principles for water management;

   e) Regulates the general water use regime and its priorities;

   f) Defines the rights and obligations of water users

g) Determines the institutional organization of the water sector.

The National Territorial Development Plan (PNDT) was recently considered by the Council of Ministers and submitted to Parliament in 2020 for approval. It recognizes the importance of flood risk and the need to integrate flood risk assessment and management into the planning process, in order to implement the policy of minimizing potential flood risk; it is part of a range of responses to this type of event including risk assessment, forecasting and early warning of flood events, emergency response systems and other hydraulic engineering projects. Proper planning in the context of sustainable development can, in exceptional circumstances, require incremental modifications in areas at risk, provided it is managed properly;

National Strategy for Adaptation and Mitigation of Climate Change (ENAMMC, 2013). ENAMMC recognizes that climate change is a dynamic process that requires a preventive and flexible approach to ensure adequate intervention, promoting adaptation to minimize its potential consequences;

National Development Strategy (END, 2014). It recognizes that the implications for biodiversity must be considered at all stages of assessment and management.

1.11.1 Occupation and use of space

Occupation: form of acquisition of the right to use and enjoy the land by natural persons who have been using the land in good faith for at least ten years, or by local communities. (BR - Law no. 19/97 - Legislation on Land).

The complexity formed by urban space and the city cannot do without well-defined, structured and comprehensive urban planning that, in a multi-sectoral way, takes into account man and the environment in economic, social, physical-territorial, ecological and administrative aspects. In general terms, the set of different land uses juxtaposed to each other. These uses define areas such as: The city center, the place where commercial, service and management activities are concentrated; industrial areas and residential areas, distinct in terms of form and social content; leisure areas; and, among others, those set aside for future expansion. This set of land uses is the spatial organization of the city or simply fragmented urban space (Santos, 2010).

Human settlements and the way they are organized in space are dynamic phenomena that acquire their own characteristics according to the type of social, economic and cultural development of the human group and the way they relate to the land (Araújo, 1997).

Territorial planning is established to ensure the organization of the national space and the sustainable use of its natural resources, observing the legal, administrative, cultural and material conditions favourable to the country's economic development, the promotion of people's quality of life and the protection and conservation of the environment (MICOA, 2009).

Law no. 19/2007 on Territorial Planning in Mozambique, approved by the Assembly of the Republic on July 18, in Article 13, no. 2, states that the preparation of Territorial Planning Instruments (IOT) is mandatory for the District and Municipal levels. Thus, the need to draw up these instruments (Urban Structure Plan, General Spatial Planning Plans and Detailed Plans) described in Article 10, number five, becomes imperative, as a way of promoting harmonious and balanced economic and social development, both in urban centers and in rural areas.

Each of the above-mentioned land-use planning instruments, for example the detailed plans, define in detail the type of occupation of any specific area of the urban center, establishing the design of the urban space, providing for land uses and general building conditions, the layout of roads, the characteristics of infrastructure networks and services, both for new areas and for existing areas, characterizing the façades of buildings and the layout of open spaces.

According to Narvais (2011), land occupation is the possession and use of land for productive activities (e.g. agriculture and commerce) or for the settlement of people (e.g. building houses).

1.11.2 Informal occupation

Informal occupation refers to any area that has arisen through an illegal process, consisting initially of improvised housing with a precarious construction standard and urban infrastructure problems, regardless of whether it was illegally built on third-party land and may also be located in central areas of the city.

Costa & Rocha (2010) state that informal occupations are a reality all over the world and present precarious socio-environmental and socio-economic conditions, influencing the health of their residents and the environment in which they live. As such, it has its origins in:

a) The inability of the low-income population to compete in the capitalist housing market, leading to exclusion;

b) The need to reduce housing costs in the face of low wages (underemployment) or unemployment;

c) Rural migration, caused by rural economic stagnation, the modernization of agricultural activities and the attraction of industrial expansion in the cities.

The growing urbanization process and the complexity of the problems arising from it point to a major challenge for the sustainable development of cities, especially in terms of the organization of urban space. Among the factors hampering the harmonious development of the city is the process of informal occupation. This process leads to an excessive extension of the urban fabric with low densities of occupation, making it impossible to implement basic infrastructure and services.

Informal occupation is promoted at the initiative of the population by encroaching on public land and areas of partial protection. As well as contributing to functional disintegration, informal occupation hinders planning and the implementation of road links, the location of community facilities and the infrastructure network in neighborhoods and subdivisions.

The occupation and development of living spaces, be they in the countryside or in the city, cannot take place haphazardly, but in accordance with private and collective interests. Studies are needed into the nature of the occupation, its purpose, an assessment of the local geography, the capacity to support this use without damaging the environment, in order to provide good living conditions for people, enabling economic and social development that harmonizes private interests and those of the community (Caomba, 2018).

Informal land occupation exposes a variety of problems. Araújo (2003) analyzes these problems, whether due to inadequate planning, lack of planning or omission on the part of public authorities, and defines them as a result:

a) **Alteration of the production regime**: soil sealing prevents water infiltration, accentuating urban erosion problems and increasing flood peaks, minimizing soil recharge reduces water availability during periods of low rainfall.

b) **Lack of basic infrastructure**: the lack of sewage collection and treatment and the inadequate disposal of waste bring contaminants to the rivers, which have compromised water quality, making it difficult to purify the water.

c) **Waste**: Different uses of water, coupled with its apparent abundance, lead to negligent use and mismanagement of this natural resource, resulting in significant waste.

1.11.3 Factors in informal employment

The process of urbanization is the result of an increase in population density on the outskirts of concrete areas, resulting in an overload of basic services and infrastructures. It can be clearly seen that informal occupations are more often than not responsible for various problems linked to the social and environmental aspects that arise specifically from the poor management of urban space, hence the need for good planning of housing space to house various services that can help improve the quality of life of the resident population itself (MICOA, 2010). These have the following characteristics:

a) Disorderly and spontaneous occupation of physical space;
b) Physical occupation without defined and sufficient living space;
c) No safe and durable housing;
d) Poor access roads;
e) Insufficient access to basic sanitation;
f) Insufficient access to drinking water;
g) Lack of basic health care;
h) High rates of poverty and unemployment.

1.11.4 Socio-environmental impacts of informal occupation of space

According to Ribeiro (1999: 34), the informal occupation of space causes socio-environmental impacts on the geophysical environment such as:

a) Rain erosion;
b) Soil compaction;
c) Change in topography;
d) Soil impermeability;
e) Decreased infiltration;
f) Increased surface and underground runoff;
g) Microclimate change.

Understanding space as a social production leads to the realization that socio-environmental issues stem from the relationship between man and nature. The environmental issue refers not only to the problems of nature, the physical environment, but also to the problems arising from societal action (Rodrigues, 1986).

CHAPTER 2: METHODOLOGY

The research was carried out in the following stages:

Figure7 : Methodological overview

Source: Author, 2024.

In order to achieve the objectives set, a descriptive and prospective study was chosen, adopting a mixed approach that combines quantitative and qualitative methods. The choice of the descriptive approach is justified by the interest in detailing the acceptance and perception of risks by the residents of downtown Ricatla during the process of occupying spaces in an area known to be vulnerable to flooding.

The qualitative approach was selected in order to understand the social aspects underlying the phenomenon under study. This approach is based on the analysis of the discussion and correlation of interpersonal data, the direct participation of informants in relevant situations, and the interpretation of the meanings attributed by them to their actions.

As for the quantitative approach, it was adopted in order to lend objectivity to the research. The data collected was interpreted using statistical tools, including measures of central tendency (mean, mode and median) and measures of dispersion (variance and standard deviation).

In this study, we highlight the use of the chi-squared statistical test as an instrument to validate the hypothesis formulated. This specific test is relevant for assessing the relationship between categorical variables, adding statistical robustness to the research. In this way, the combination of descriptive, prospective, qualitative and quantitative methods provides a comprehensive approach, allowing for a deeper and more meaningful analysis of the acceptance and perception of risks by the residents of downtown Ricatla in an area prone to flooding.

2.1 Data collection technique

In terms of data collection techniques, we worked with bibliographical and documentary research, interviews and questionnaires (the latter used closed and open questions to obtain sufficient data to answer the research problem).

2.1. 1. Procedures

The work began with a bibliographical and documentary review. The bibliographic review is fundamentally based on the contributions of various authors and this method aims to understand everything that has been written about the perception and acceptability of flood risk. This involved collecting geological data from the National Geology Directorate (DNG), the Mozambique Agricultural Research Institute (IIAM), the Ministry of Land and Environment (MITA), the Provincial Land and Environment Directorate, the National Disaster Management Institute (INGD), the Geology Department Library (DG), the Brazão Mazula library at Eduardo Mondlane University (UEM), and the library at Maputo Pedagogical University (UPM).

While documentary research makes use of material that has not yet received analytical treatment, or which can still be reworked according to the research objectives. Documents can be found in physical and electronic format. Physical documents are: official documents, newspaper reports , letters, diaries, films, photographs. Electronic documents are research reports, company reports, statistical tables and others. In the context of this research, we would highlight the use of reports, newspapers, magazines and photographs from various institutions such as:

a) Law no. 19/1997 of October 1 - Land Law;
b) Law no. 19/2007 of July 18 - Spatial Planning Law;
c) Decree no. 60/2006 of December 26 - Urban Land Regulations;
d) Law no. 20/1997 of October 1 - Environmental Law;
e) Law no. 10/2020 of August 14 - Disaster Risk Reduction and Management Act; and
f) Law no. 15/2014 of June 20 - Establishes the Legal Framework for Disaster Management.

The second stage involved fieldwork based on semi-structured interviews with 100 people using a questionnaire with open and closed questions relating to the identification of the socio-economic profile (for the interviewee's perception and experience of the phenomenon studied), the identification of the subject's perception of the causality and responsibility for the floods, the residents' assessment and choice of place to live, the frequency of leaving and returning to the area and the action and participation of the district government in the community's problems.

The heads of households were interviewed, as were the members of the same household vulnerable to risk, regardless of their age or gender. In order to gather different opinions on the acceptability of risk, we also interviewed community leaders and institutions involved in land and risk management in Marracuene, such as the District Planning and Infrastructure Services (SDPI) and the Risk Management Committee through their representatives, and we also heard the opinion of an environmentalist and a physical planner.

It should be noted that the study area has serious mobility and accessibility problems, which to some extent contributed to the length of time it took to meet the interviewees. The presence of the community leader during the interview phase was crucial, as the community leader facilitated access to the area for those involved in the research in an attempt to minimize any possible conflict caused by a person from outside the community, thus providing greater security for those involved. This is because, for the residents of lower Ricatla, the researchers don't bring solutions to their problems or help the people who are currently carrying their belongings in the water.

The third step was to use the cartographic method to produce a location map and thematic maps (hypsometer map, soil map and infrastructure map), using ArcGIS software version 10.8 and *SASPlanet* to visualize the study area. The intervention area was delimited on the basis of the aerial image. The study area, Baixa do Ricatla in the district of Marracuene, was selected on the basis of specific criteria aimed at understanding the acceptability of the flood risk.

GPS was used to remove points representing the geographical limits of the Baixa, which allowed for precise and objective delimitation, guaranteeing a faithful representation of the area of interest. By using geospatial data, it was possible to identify and analyze the physical and social characteristics of Baixa de Ricatla, including its proximity to bodies of water, population density, vulnerability of existing structures and proximity criteria.

The hypsometric map was generated from the Digital Elevation Model (DEM) with a spatial resolution of 12.5m. It is available free of charge at: https://asf.alaska.edu for ALOS PALSAR. Once the DEM was available, imperfections were corrected using the Fill tool in the *Spatial analysis tool* in arcmap after classification into 5 classes. To produce the soil map, the soil map of Maputo province was vectorized at a scale of 1:50 000.

2.2 Population and sample

The population of this study was made up of residents of downtown Ricatla who were available to live in the vicinity of downtown at the time of the survey, since most of them were sheltering in the homes of relatives and close friends as their homes were still flooded at the time of data collection due to the rainfall of the previous rainy season. Data provided by the neighborhood secretary indicated that there were around 215 families in downtown Ricatla at the time.

To calculate the sample size, a significance level of 95% was established with a margin of error of 5% of the households. The sample for this study consisted of 120 families and was determined as follows:

$$n = \frac{z^2 x\, p\, (1-p) x\, N}{d^2 x\, (N-1) + z^2 x\, p(1-P)}$$

$$n = \frac{1.96^2 x\, 0.5\, (1-0.5) x\, 215}{0.05^2 x\, (215-1) + 1.96^2 x\, 0.5(1-0.5)} \tag{1}$$

$$n = 120{,}279 \approx 121$$

Legend: n - sample size; z - confidence level (1.96); p - population proportion of individuals belonging to the category of interest (0.5); N - population size, and d - maximum permitted error (0.05).

Simple random probability sampling was used. After surveying the families available from the neighbourhood secretary, they were organized and identified using codes assigned within the scope of the survey. The participating families were selected at random to ensure that all samples had the same probability of taking part in the survey. In order to better understand the selection of the sample used in this research, Figure 8 shows the scheme for determining the sample.

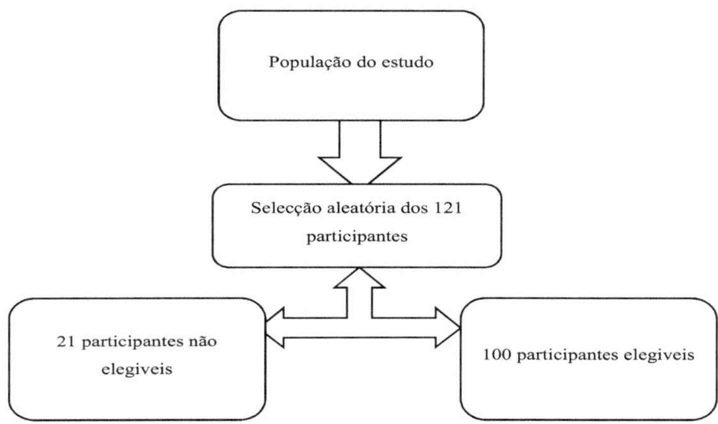

Source: Author, 2024.

2.2.1 Profile of interviewees

The following tables illustrate the characteristics of the residents interviewed in Ricatla in terms of gender, age, education and marital status.

With regard to the socio-demographic variables of the participants, in summary, of the 100 interviewees, 60% of the households were male and 40% of the households were female. With regard to age, the majority of participants, 56% of households, were aged between 18-24, followed by 28% of households in the 25-34 age bracket. The lowest frequency was seen in the over 55s group, with only 3% of households.

Figure9 : Gender and age distribution of interviewees

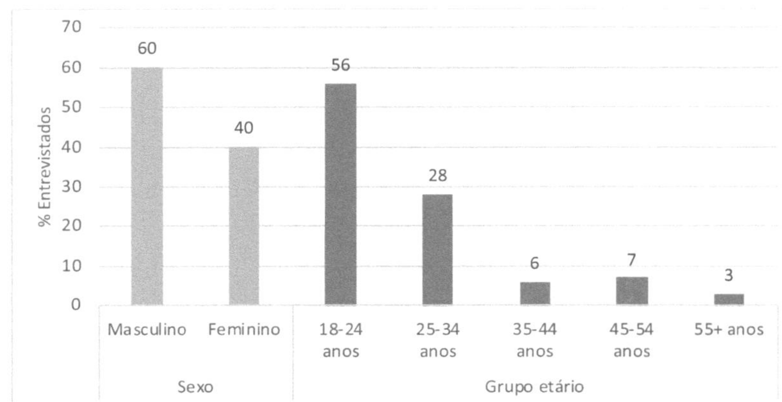

Source: Author, 2024.

These socio-demographic characteristics show the panorama of Ricatla's population. The predominance of young adults (56% between the ages of 18-24) may influence the needs and priorities of the place, including demands for education, employment and services aimed at this age group.

Gender disparity, with a 60% male majority of household heads, can also have significant implications for the social and economic dynamics of the place. Schooling, with 40% of household heads having only a primary level education, highlights the need for investment in education to improve opportunities for personal and professional development.

Figure10 : Level of education and marital status of interviewees

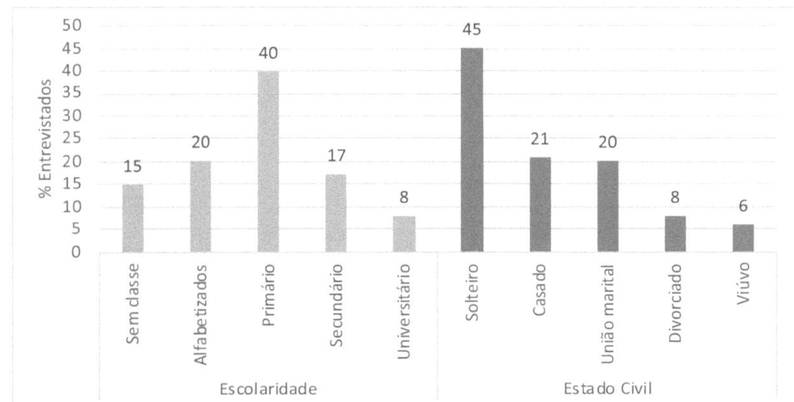

Source: Author, 2024.

As for the level of education, the majority of participants (40%) reported that they had completed primary school. Another 20% of households were literate, and 8% of households had completed university education (Table 8).

The marital status of the heads of household shows the diversity, with 45% being single and a considerable proportion of households (20%) headed by people living in a marital union. This may reflect different family structures and social support needs

2.3 Inclusion and exclusion criteria

To have lived in downtown Ricatla for at least one year, to be close to the risk zone and to agree to take part in the research by signing a free informed consent form.

Fifteen families were excluded from the survey because they had been part of the research and had moved in search of shelter, making it impossible to actually establish contact for the interview, and another six families whose representatives were not available to answer the questionnaire used for data collection.

2.4 Variables

The variables in this study were separated into three categories according to the research objectives.

Table4 : Research variables

Variables		
Sociodemographi **c**	**Acceptability**	**Risk perception**
Gender, age, profession, education, marital status	When you think there is flooding, the possibility of water entering your home, flood warning, control of the risk of flooding, attitude towards the warning, having already left the place, reason for returning after flooding, neighborhood initiatives for minimization, having already benefited from resettlement and whether you would leave the place for a safe place.	Reason for living in downtown Ricatla, awareness of flood risk, when floods occur, flood damage, degree of flood risk, water level during floods, degree of information and causes of floods

Source: Author, 2024.

2.5. Data analysis and interpretation

The information gathered during the research was coded, organized, systematized, grouped into tables and graphs to provide greater clarity and guarantee its integrity. After collecting the data, it was compiled in an Excel spreadsheet and exported to the *Statiscal productland specific software* (SPSS) platform where it was analyzed using descriptive statistics tools, specifically measures of central tendency (mean, median and mode), measures of dispersion (standard deviation and variance) and the chi-squared test ($x^{2)}$ to verify the hypotheses. The significance level considered was 5%, which was previously established when calculating the sample size.

2.6 Ethical considerations

In order to carry out this work, some ethical research principles were observed, and authorization was requested from the study site, starting with the Marracuene District and ending in the Ricatla lowlands.

Thus, the participants in this study had the right to accept or refuse to take part in the work, as well as the right to discontinue their participation at any time they deemed necessary without being affected by the decision (autonomy), even considering that taking part in this work would not bring them any material benefit, the residents of the Ricatla neighborhood were made aware of the importance of knowing at national and international level the degree of risk, as well as the situation of vulnerability in which they find themselves (beneficence).

The collection of information consisted of an interview without physical contact or submission to any type of risky action (non-maleficence). During the work, all participants were treated in the same way regardless of social level, education, religion, origin, etc. (fairness and equality). All the participants' personal information was safeguarded between the principal investigator and the tutor (confidentiality and secrecy). The interview lasted just 10 minutes, in a private place (privacy). For identification purposes, codes assigned by the researcher were used (anonymity).

CHAPTER 3: GENERAL DESCRIPTION OF THE STUDY AREA

This chapter deals with the physical-geographical and social characteristics of the study area, based on bibliographical consultations, analysis and the compilation of information gathered in the field.

3.1 Geographical location of Baixa do Ricatla

The study area, Baixa de Ricatla, covers a total area of 4.8 km² and is located in the District of Marracuene, Administrative Post of Marracuene-Sede, in the Locality of Michafutene, situated in the eastern part of Maputo Province, approximately 30 km north of Maputo City (Simbine, 2023).

Geographically, Baixa de Ricatla is located in the southwestern part of the Marracuene District, covering the Ricatla, Zintava and Agostinho Neto neighborhoods, between latitudes 25° 45' 40" to 25° 48' 00" South and longitudes 32° 36' 30" to 32° 38' 00" East, (see figure 11).

Figure11 : Geographical location map of the area

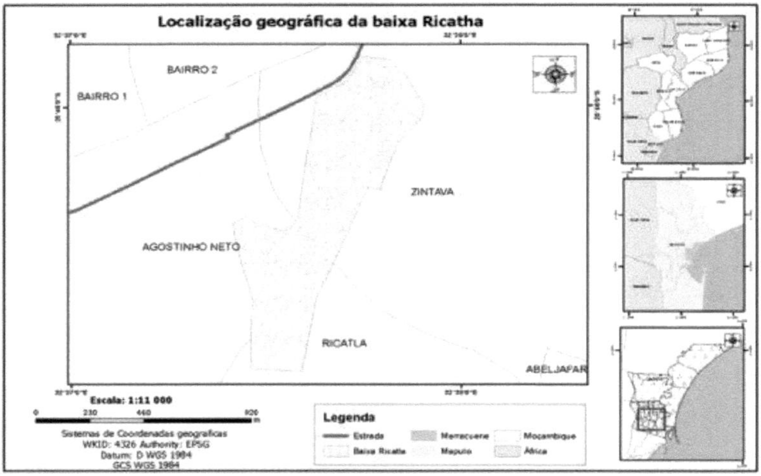

Source: Author, 2024

3.2 Physical and Natural Characterization

Physical-natural analysis is an important component in understanding the vulnerability and risks associated with natural events such as floods and assessing the acceptability of flood risk. This section addresses the physical-natural aspects, which directly influence susceptibility to natural disasters, especially considering topography and soil composition that can affect drainage and water retention.

3.2.1 Geology

In terms of geological structure, the Ricatla Lowlands belong to the Phanerozoic and are made up of the Ponta Vermelha Formations, dating from the upper Pliocene, and are made up of sedimentary rocks from the Tertiary Period characterized by the occurrence of Aeolian sand, siltstone and red sandstone, and are also made up of sedimentary rocks from the Quaternary Period characterized by the occurrence of alluvium, sand, silt, gravel; inland dune; red Aeolian sand and are from the Quarternary Period (Geological Chart, scale 1:250 000, n°2532).

3.2.2 Relief

The Marracuene district has a coastal strip of sand dunes, separating the sea from the Incomáti River in the Macaneta area, which is at risk of disappearing, which would have serious ecological consequences for the Marracuene, Manhiça and Magude districts. The upper zone of the district is mainly made up of aeolian sandy sediments, to the west and along the coast, with the occurrence of silent areas (MAE, 2014).

In topographical terms, the town of Ricatla is located in the Great Plains area, with altitudes varying between 20 and 44 meters above sea level. The relief is predominantly flat with a few small elevations. It is steep in its entirety, with minimum heights ranging from 20 to 30 meters in the central part of the Baixa, and maximum heights ranging from 39 to 43 meters in the southwest of the Baixa de Ricatla (*Figure 12*).

Given the configuration of the relief and the behavior of the contour lines in the region, it is clear that Lower Ricatla has lower altitude values compared to all the surrounding directions (north, south, east and west). The altitudes in this area vary from 20 to 43 meters, making it a natural water accumulation point. This peculiar topography makes Lower Ricatla a kind of catchment basin, where water converges and accumulates easily, especially during periods of heavy rain.

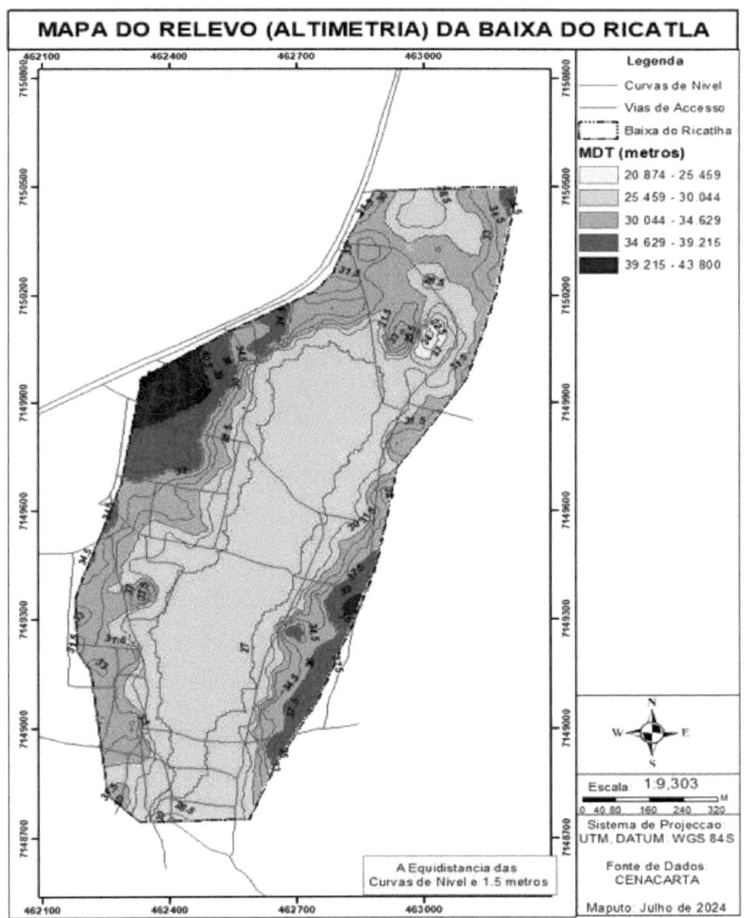

Source: Author, 2024.

The concentration of water in this depression results in frequent flooding, seriously affecting the infrastructure and quality of life of the neighborhood's residents. Flooded streets, damage to property and disruption to public services are just some of the challenges residents face. Frequent flooding increases the risk of waterborne diseases and causes significant environmental damage.

Figure13 : Low-lying and flooded areas

Caption: *Submerged houses in downtown Ricatla.* Source: Author, 2024.

3.2.3 Soils

The Ricatla lowlands are characterized by the predominance of sandy soils, which have well-distributed sand particles, resulting in a loose and porous structure. These soils have high permeability, which allows for rapid water infiltration, but also means that they have low moisture retention capacity. This characteristic can be problematic in an area prone to flooding, as is the case in Baixa deRicatla.

According to PEUCM (2010), from a geological point of view, these soils correspond to eolian sands from the upper Pleistocene. Topographically, they are almost flat soils, with a slope ranging from 0 to 2%. The soil texture on the surface is predominantly sandy to sandy-white, while in the subsoil it is sandy. The depth of the soil can be more than 1 meter, with drainage ranging from imperfect to moderate. During periods of heavy rainfall, rapid infiltration can saturate the soil quickly, while the lack of retention capacity can aggravate surface runoff, increasing the risk of flooding. In addition, the fragile structure of sandy soils can be easily eroded, especially in areas where vegetation is scarce or has been removed. These conditions complicate urban planning and environmental management in the region, requiring specific solutions for mitigating the impacts of flooding and soil conservation.

Figure14 : Soil Map of Lower Ricatla

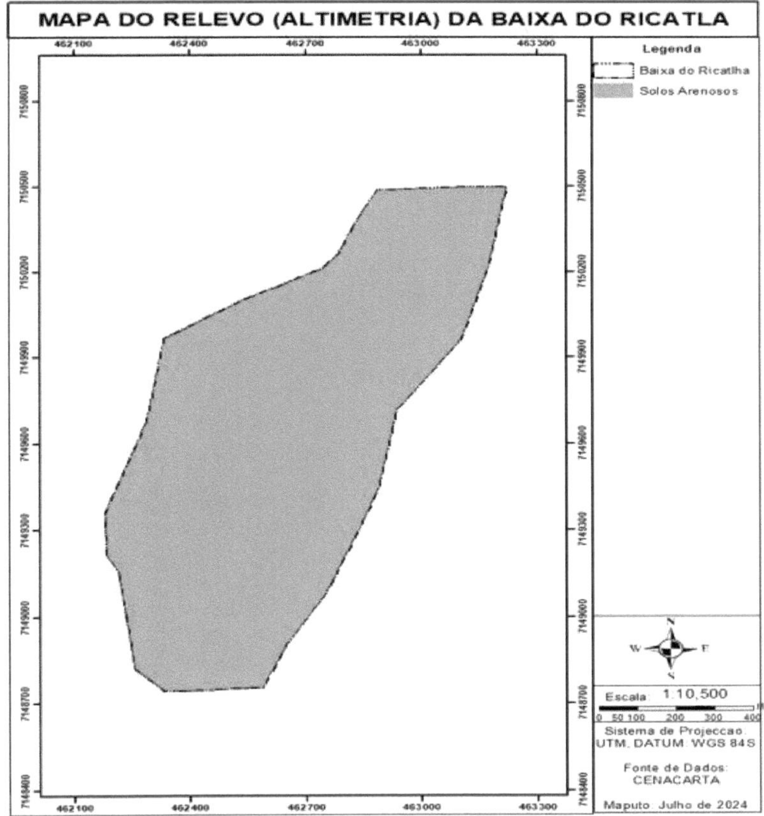

Source: Author, 2024.

3.2.4 Climate

According to Muchangos (1999), the climate of southern Mozambique is clearly tropical. The greatest influence on the climate of this region, and particularly with regard to the behavior of rainfall and temperature, is exerted by the location in the area of the south-east trade winds, by the Mozambique sea current.

Lower Ricatla is located in a region whose climate is characterized by alternations between dry conditions, induced by high subcontinental pressure and the incursions of humid winds from the ocean, and cold waves that bring violent storms. According to the Koppen classification, Marracuene has a humid tropical climate, with two (2) predominant seasons: the hot, high rainfall season from October to March; and the cool, dry season - from April to September. The average annual rainfall is 141 mm, concentrated in the months of December to February, and the average annual temperature is 23.1 °C (*see graph in figure 15*).

Figure15 : Thermopluviometric graph

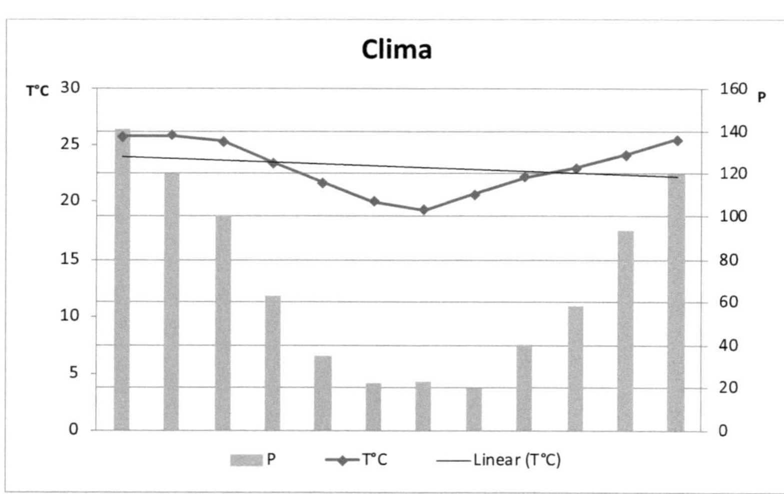

Source: INAM data.

3.2.5 Vegetation

In Baixa de Ricatla, the vegetation cover is made up of native herbaceous and submerged anchored vegetation that is tending to disappear. There are species adapted to local conditions whose scientific names are shown in Table 9, such as: Mafurreiras, Mangueira, Coqueiro, Cajueiro, Abacateira and Canhueiro. However, a considerable part of the natural vegetation has been profoundly altered by human influence, particularly due to the felling of trees for the construction of houses, exploitation for building material and wood fuel.

Table5 : Index of plant names

CURRENT NAME	SCIENTIFIC NAME
Cashew	Anarcadium occidentale L.
Kidney beans	Vigna unguiculata
Moringa (various species)	Moringa spp.
Avocado	Persea americana Mill.
Mafureira	Trichilia emitica
Hose	Mangifera indica L.
Canhoeiro	Sclerocarya birrea subsp.caffa
Coconut tree	Cocos nucifera L.

Source: Author, 2024.

3.3 Characteristics socioeconomics

3.3.1 Population of Ricatla

The town of Ricatla belongs to the administrative post of Marracuene-Sede, which is considered to be the second most populous post in the district. The administrative post has a total of 20,372 inhabitants, of which 9,652 are male and 10,720 are female (INE, 2017).

It should be noted that, according to the risk management committee, downtown Ricatla has a total of 759 families, corresponding to 3795 inhabitants. Most of these are young and were born in the neighborhood, while others have taken over spaces from locals. It can be seen that there are fewer elderly people, which may be a consequence of the resettlement that took place four years ago. However, some of the resettled people have returned to the study area leaving their older children behind, in many cases, and in others they have sold their previous homes to new residents with the complacency of the local authorities, thus justifying the number of young people in the area.

The size (average of 5 members per family) and age structure of households (more young people than adults) and the high illiteracy rates, especially among women, which are reflected in the high poverty incidence rate (53%), are important factors in the high exposure (due to the occupation of susceptible areas) and propensity (such as reduced capacity for anticipation and resilience) of a large part of the country's population to natural risks (Coamba, 2023).

3.3.2 Economic activities

The natural resource potential that the lower Ricatla offers is the proximity of rocks that generate fertile land and, as it is a flood plain, it is very suitable for farming. There is currently a portion of land that has not been occupied by houses, on which the population (natives and some residents of adjacent neighborhoods) practice subsistence farming.

Figure16 : Agricultural activity in Lower Ricatla

Caption: A photograph of crops grown in downtown Ricatla. Source: Author, 2024
According to one of the block leaders, in addition to subsistence farming, the native population, especially the young teenagers, have taken up fishing, as we can see in figure 17.

Figure17 : Fishing in downtown Ricatla

Caption: a) Photograph shows fish being caught. b) Illustrates the young man who caught fish in downtown Ricatla for later sale. Source: Author, 2024.

3.4 Infrastructure

According to the information obtained, the study area has an informal market, which has a total approximately 30 wooden stalls. The residents' source of income is diverse because some depend on informal jobs, others on formal jobs and some work in commerce.

Ricatla's economic base is made up of a range of income-generating activities capable of maintaining and developing its physical and social environment, such as beauty salons, carpentry, metalwork, poultry farming, fishing and informal shoemaking, which is the sector that occupies the most people.

3.4.1 Housing

According to Corona and Lemos (1972), housing is the place where people live and reside. In the field of architecture, it is seen as the shelter or enclosure that protects human beings, providing adequate conditions for their existence in both material and spiritual terms. Housing provides a safe and comfortable space that encourages daily activities and emotional well-being, and is often compared to a dwelling or residence. In addition to the function of protection from natural elements, housing plays a crucial role in shaping individuals' identity and sense of belonging, reflecting cultural, social and personal values.

According to Albernaz and Lima (2000), housing is the built space used for living in. It can be single-family, when it is intended for a single family, and is commonly called a house, or multi-family, when it is intended for more than one household, such as an apartment building.

The lower part Ricatla has conventional buildings and some precarious ones. The streets are very narrow (alleys), without any subdivisions, which makes life difficult for residents in terms of mobility and accessibility of people and goods.

Figure18 : Flooded houses

Caption: a) House of precarious construction flooded. b) House of conventional construction flooded: Source: Author, 2024

It should be noted that none of the residents had been allocated space by the Marracuene district planning and infrastructure service (SDPI) and that the occupations were spontaneous. With regard to legal documentation for the use and exploitation of land, the results show that 90% of our households do not have a DUAT (Right to Use and Exploit Land), only a small proportion of households do, so only 10% do.

According to the SPDI in Marracuene, the Baixa de Ricatla is an area in which it is forbidden to issue a DUAT for any type of use in that space, the government refusing to process the process because it is an area unfit for habitation. The picture below shows houses submerged by rainwater. Even so, most of the population has not abandoned their homes.

Figure19 : House submerged by rainwater

Caption: The photograph shows flooded houses, with the water reaching the height of the general beam in Lower Ricatla. Source: Author, 2024.

3.4.2 Road network

As far as the road network is concerned, downtown Ricatla is crossed by National Road No·1 (EN1), which plays a key role in the distribution of the tertiary roads that connect it internally. There is asphalt only on National Road No. 1, which connects the south and north of the country, passing through downtown Ricatla. The footpaths are mainly found in unplanned residential areas, such as the settlements located on the flood plains (figure 16).

Figure20 : Crossings built over a watercourse

a) The photograph shows a makeshift bridge over a stream/drainage ditch to facilitate crossing the watercourse. b) It illustrates the mechanism adopted for crossing, using sandbags on a flood plain. Source: Author, 2024.

3.4.3 Power grid

As for the electricity grid, the Lower Town of Ricatla is supplied by the national electricity grid, and most of the homes have it. However, in the informal settlement areas, there is no access to the electricity distribution network due to their territorial organization.

3.4.4 Water supply network

With regard to water supply, the area under study has a public piped water supply system, Small Water Supply Systems (PSAA) and groundwater wells. The neighborhood has a piped water supply from the Water Assets Investment Fund (FIPAG) and PSAAs. However, there are families who use wells built in their homes to guarantee access to water (see Figure 21)

Caption: a) Piped water supply system. b) GIMOC's Autonomous Sanitation Service Point (PSSA). c) Groundwater wells used for consumption in Lower Ricatla. Source: Author, 2024.

CHAPTER 4: RESULTS AND DISCUSSIONS

The presentation and analysis of data is the last stage of this research, where we try to reconstruct the analyzed data as a structured and meaningful whole, explaining the research products and the interpretation of them, taking into account the research objectives.

4.1 Types of infrastructure

This chapter describes the type of housing, energy and water sources, access to education, health and land used by the residents of Ricatla. It helps us understand the type of infrastructure in place, bearing in mind that this is an area at risk of flooding.

4.1.1 Housing

With regard to the material used to build their homes, 40% of households said they lived in block houses, 30% in wood and zinc houses, 20% in reed houses, 6% in brick houses and 4% in matope houses. These figures indicate a predominance of less durable materials such as wood, zinc and reeds, which may reflect structural and economic vulnerabilities, as shown in figure 22.

Figure22 : Material used to build houses as a percentage of households

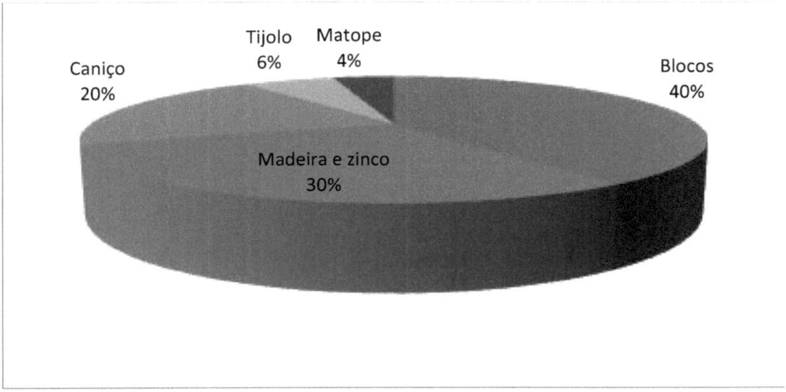

Source: Author 2024

4.1.2 Access to water

As for the source of water, 52% used boreholes, 32% wells and 16% supply systems. The predominance of the use of wells shows dependence on water sources that are less secure and more susceptible to contamination.

Figure23 : Water sources used by residents in Percentage of households

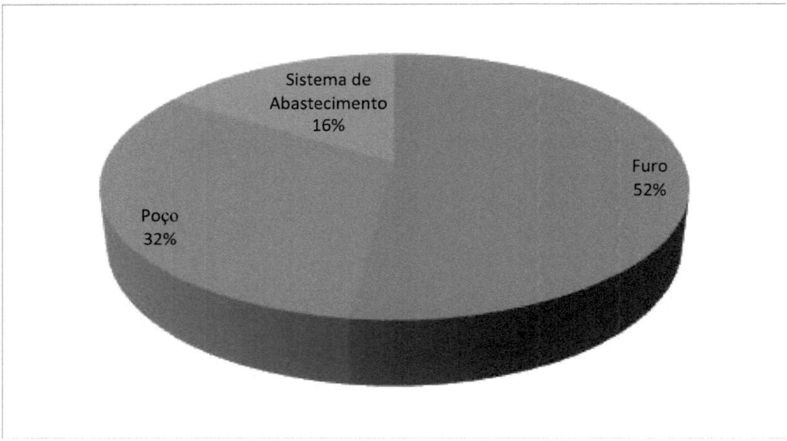

Source: Author 2024

4.1.2 Access to energy

This graph shows the different energy sources used for domestic use in the kitchen and for lighting. As this is a high-risk area, it would be normal for there to be no power lines because of the danger to residents. The data shows that 37% of households use coal for cooking, 20% use oil, 19% firewood, 16% electricity, and 1% use gas. Those who don't have electricity use candles as a source of lighting, which account for 7%, as shown in figure 24. The high dependence on coal and oil as energy sources indicates possible environmental impacts and health problems associated with burning these fuels and reflects the socio-economic situation with low purchasing power.

Figure24 : Energy sources used by residents in Percentage of households

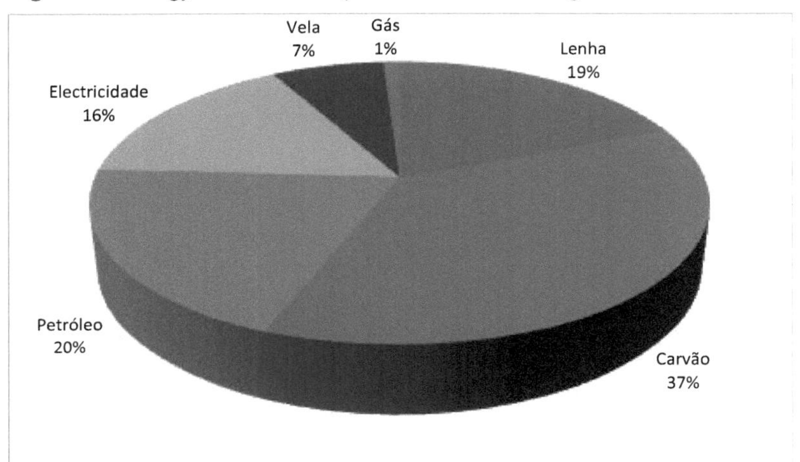

Source: Author 2024

4.1.3 Access to healthcare

With regard to access to health services, 86% of households say they have access to health services, while 14% say they don't have access to these services. In this survey, it is understood that the negative response regarding access to services demonstrates a refusal to use the services, because they are available in the study area. This high percentage of access to health services is positive, but there is still a significant portion of the population that needs improvements in access to medical care.

Figure25 : Access to health services as a percentage of households

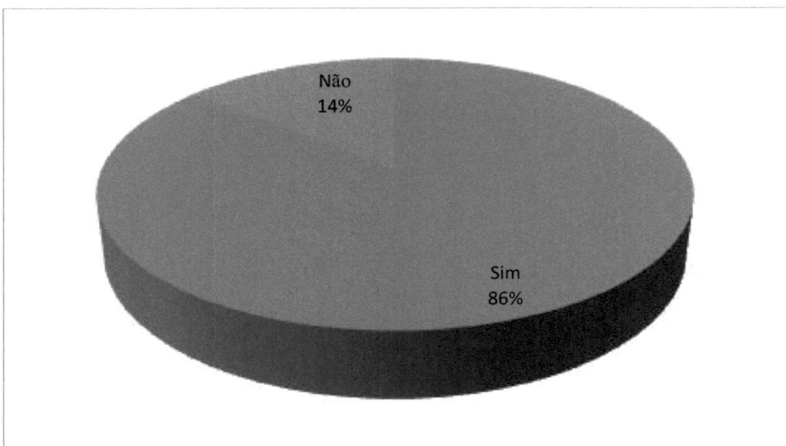

Source: Author 2024

4.1.4 Access to educational infrastructure by level

The graph in Figure 26 shows the distribution of existing educational infrastructures downtown Ricatla at their different levels of education, where 64% say they have access to EP2, EP1 accounts for 16%, ESG1 with 12% ESG2 and the remaining 7% and 1% of those representing the children's centers in the area.

Figure26 : Graph of educational infrastructures in percentages.

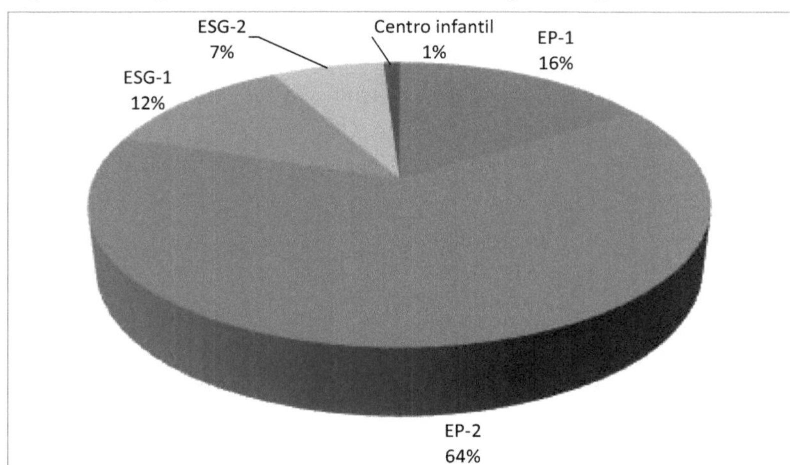

Source: Author 2024

4.1.5 Length of residence and access to land

The question about the length of residence in the study site arose as a way of understanding how long the interviewees who took part in this survey had been living with the risk in downtown Ricatla. It is possible to see that we have a greater representation of residents who have lived in the area for less than 5 years, which represents 50% of our households, 21% say they have lived in the area for between 5 and 10 years and the remainder, which corresponds to 29% of households, for more than 10 years.

While the graph in figure 27 shows the different ways of acquiring space, the latter leads us to think that it is the natives who are also part of those who leave in the rainy season and return as soon as the rainy season is over.

The majority of residents, which corresponds to 70% of the community, bought their plots from locals, even though this is an illegal act, 28% received them as inheritance and 2% in good faith. The data referring to the purchase of space justifies the position of the district government, which claims that it was not part of the allocation of space downtown and that the population did so illegally.

Figure27 : Length of residence on the plot and how the plot was acquired.

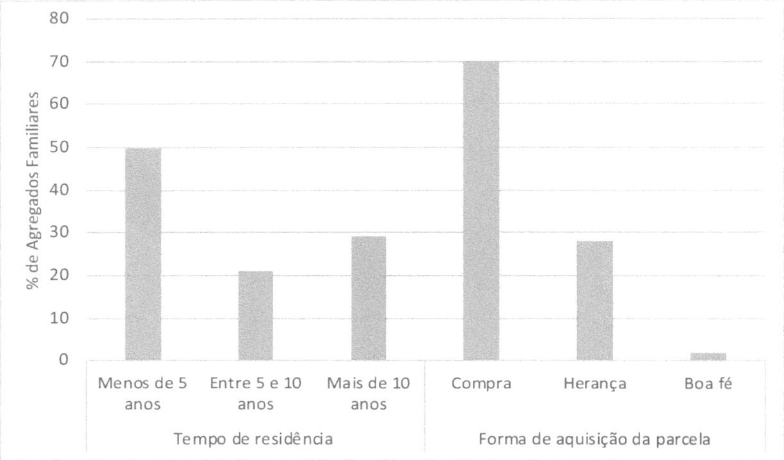

Source: Author, 2024.

4.2 Parameters of the perception of flooding in lower Ricatla

In order to better understand the degree of flood risk perception in downtown Ricatla, a number of parameters were used, such as: perception of flood risk, reason for construction, degree of flooding, water level, cause of flooding and alert attitudes.

4.2.1 Perception of flood risk

With regard to the perception of flood risk, 87% are aware of the risk of flooding, indicating a high level of awareness of the danger among residents. However, this perception does not translate into effective preventive action for all, possibly due to financial limitations or lack of housing alternatives, and 13% of those who do not have the perception.

4.2.2 Reason for building in the area.

The graph in figure 28 shows the reasons for building in downtown Ricatla, in which 44% of households claimed to have built because of the poor financial situation, 31% of our households claim a lack of choice and 25% are natives of the area, as the graph in figure 28 illustrates. The data reveals that the majority of residents face economic constraints that lead them to settle in vulnerable areas. In addition, a portion of the residents are natives, which reinforces the historical and emotional attachment to the place, even in the face of environmental risks.

Figure28 : Reason for building as a percentage of households

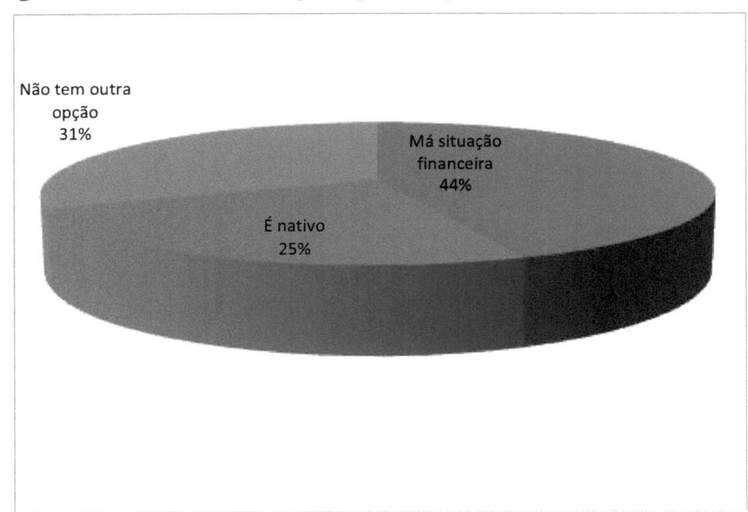

Source: Author 2024

4.2.3 Degree of flood risk

The question was asked to find out to what extent our interviewees believe they are at risk of flooding, so they assess the degree of flood risk they are at and describe it in three scenarios.

The data in the graph in figure 29 shows that 44% of households consider the degree of flooding to be high, associated with a lack of alternatives, 30% consider it to be medium and the remaining 16% consider it to be low.

financial need outweighs the perception of risk, leading them to accept living in areas vulnerable to flooding.

Figure29 : Degree of flooding as a percentage of households

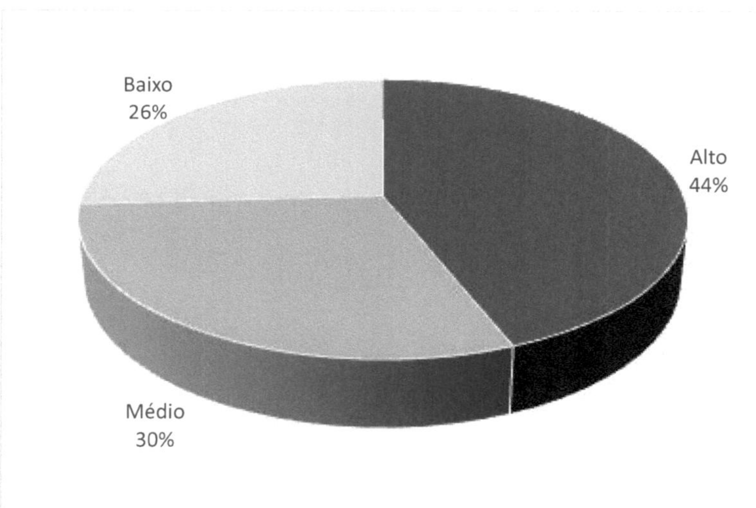

Source: Author 2024.

4.2.4 Water level

In the graph in figure 30, on the level of water during floods, 46% of our households admit that the water reaches their knees, another 40% say it reaches waist level and the remaining 14% consider that the water remains at ground level. These water levels pose challenges for the habitability and safety of homes, forcing residents to adopt improvised measures such as lifting their belongings and furniture to minimize damage.

Figure30 : Water levels during floods as a percentage of households

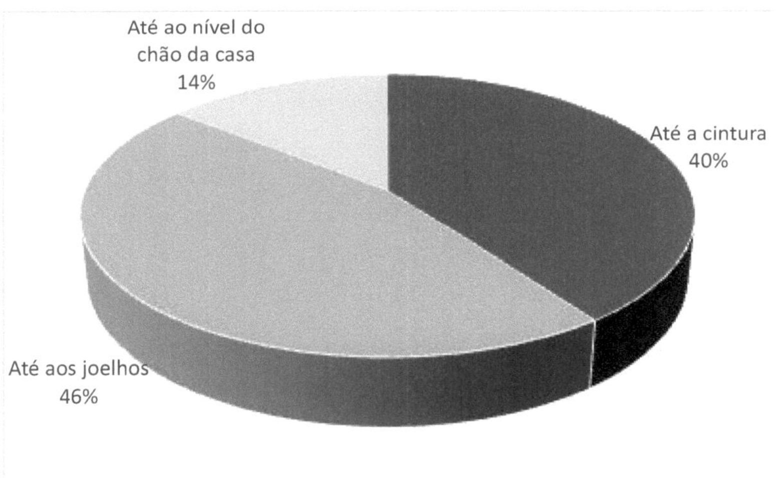

Source: Author 2024.

On the question of whether they had ever been victims of flooding, the results of the answer to this question indicate that 82% of households have already been victims of flooding, showing the recurrence of this problem in the area

4.2.5 Alert attitudes

The different attitudes of residents in response to flood warnings can be seen in the graph in figure 31. The graph shows that 50% of households prefer to leave the site, 30% of households stay while the remaining 19% adopt prevention mechanisms based on filling sandbags, using pumps to suck up water and using blocks and boxes to create elevated walkways, adapting to the flood phenomenon.

Figure31 : Residents' attitudes to flood warnings as a percentage of households

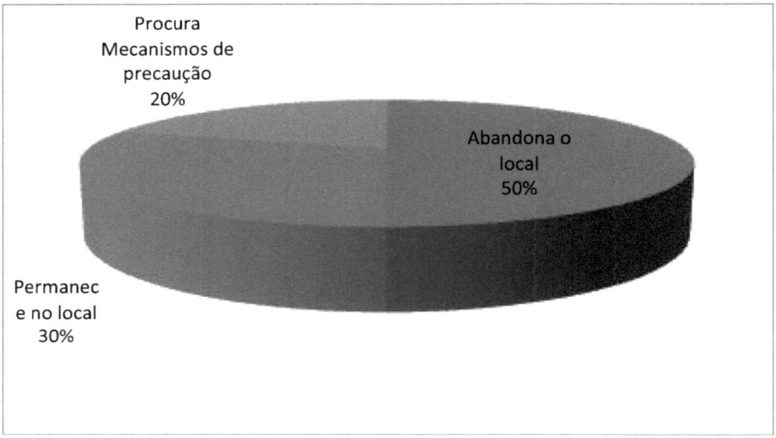

Source: Author, 2024.

These methods are corroborated by the testimonies of some residents, who show a strong connection with the place. One of the interviewees shared the following:

"This is where I was born... This is where I will die..." (Interviewee 09).

Another resident explained his adaptation strategies in more detail:

"We've managed to adapt to the rain, using boxes and blocks to lift our things inside the house. Sometimes we sleep on tables to avoid getting wet. We also set up small platforms with blocks and wood inside the house to keep our furniture and appliances above the water." (Interviewee 10).

This resilience is reinforced by another resident who says:

"I don't need to go out... As soon as the rain passes, I go back to my normal life. I also do my farming and fishing here, which is my livelihood. I'm fine here, it's close to the road and the Zimpeto wholesale market, where I can sell my goods and buy food to feed my children. Even when floods occur, we use water pumps to drain the water away from our houses and we build temporary barriers with sandbags." (Interviewee 12).

In addition to the mechanisms mentioned above, some residents mentioned the use of traditional techniques to mitigate the impacts of flooding. For example, creating makeshift drainage channels around houses to direct water to less critical areas, and building temporary retaining walls with materials available on site.

Other residents highlighted the importance of solidarity and mutual support during rainy periods. They come together to share resources, such as water pumps and sandbags, and help each other to lift their belongings and protect their homes.

Figure32 : Relationship between the degree of flooding and the water level

	Alto	Médio	Baixo
■ Até aos joelhos	61.5%	30.0%	47.7%
■ Até a cintura	34.6%	46.7%	38.6%
■ Nível do chão da casas	3.8%	23.3%	13.6%

Source: Author, 2024.

Table6 : Relationship between the reason for building downtown and the perception of risk

		Reason for building downtown					
		Financial situation		Being native		Lack of alternatives	
		N	%	N	%	N	%
Are you aware that you live in a high-risk area?	Yes	39	88.6%	24	96.0%	29	93.5%
	No	5	11.4%	1	4.0%	2	6.5%
Total		44	100.0%	25	100.0%	31	100.0%

Source: Author, 2024

Table 6 illustrates that both residents who have built in Ricatla due to lack of financial resources (88.6%), being native (96.0%) or lack of alternatives (93.5%), are aware of living in a flood risk area.

Table7 : Relationship between the reason for building in the neighborhood and the degree of flooding

		How do you rate the risk of flooding in this place?					
		High		Medium		Bass	
		N	%	N	%	N	%
Reason for building in this neighborhood	Poor financial situation	14	53.8%	10	33.3%	20	45.5%
	Being native	3	11.5%	12	40.0%	10	22.7%
	For lack of alternatives	9	34.6%	8	26.7%	14	31.8%
Total		26	100.0%	30	100.0%	44	100.0%

Source: Author, 2024.

The results of the survey indicate that there are some differences between the residents of Ricatla in terms of their knowledge of the degree of risk of flooding and the reason for building there. Of those who consider the risk to be high, the majority (53.8%) of households stated that the lack of financial conditions is the reason for building on that site. The same trend is observed among those who consider the risk to be low (45.5%). Meanwhile, for those who consider the risk of flooding to be medium, the largest proportion (40%) gave the reason of being native.

Figure33 : Possibility of water entering homes as a percentage of households

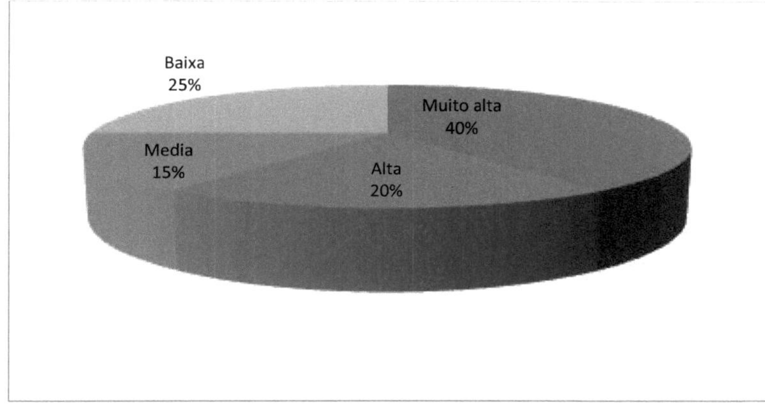

Source: Author, 2024.

In Figure 33, with regard to the possibility of water entering homes, the majority of our interviewees, representing 40%, say it is very high, 20% say it is high, another 25% say it is low and the rest say it is medium.

Table8 : Relationship between the reason for building downtown and the possibility of water entering the home

| | | \multicolumn{8}{c}{Possibility of water entering your home} |
| | | Very high | | High | | Average | | Low | |
		N	%	N	%	N	%	N	%
Reason for building in this neighbor hood	Financial situation	7	38.9%	11	44.0%	17	41.5%	9	56.3%
	Being a local	4	22.2%	6	24.0%	12	29.3%	3	18.8%
	Lack of alternativ es	7	38.9%	8	32.0%	12	29.3%	4	25.0%
Total		18	100.0%	25	100.0%	41	100.0%	16	100.0%

Source: Author, 2024.

The data shows that of those who consider the possibility of water entering their home to be low, 56.3% of households built there due to a lack of financial resources, 18.8% said they built it because it was native and the remaining 25%, due to a lack of alternatives. A similar trend is observed among those who consider the possibility of water entering their home to be high. Among those who perceive the possibility of water getting in as medium, the majority indicate the financial situation as the main reason for building in Lower Ricatla and the same percentage (29.3%) point to the fact of being native and the lack of alternatives as reasons for doing so. On the other hand, those who consider the possibility of water entering to be very high, 38.9% of households said that they built due to a lack of financial resources, the same proportion that indicated a lack of alternatives (Table 8).

Table9 : Relationship between the reason for building in the neighborhood and the effects of flooding

| | | Do you think floods cause damage | | | |
| | | Yes | | No | |
		N	%	N	%
Reason for building in this neighborhood	Financial situation	39	47.6%	5	27.8%
	Being a local	19	23.2%	6	33.3%
	Lack of alternatives	24	29.3%	7	38.9%
Total		82	100.0%	18	100.0%

Source: Author, 2024.

Table 9 shows the relationship between the reasons for building in downtown Ricatla and the perception of the effects of flooding. Among those who say that floods cause damage, a higher percentage of households (47.6%) point to the financial situation as a reason to build in downtown Ricatla; and among those who deny the damage caused by floods, the lack of alternatives is pointed out as a reason to build in the study area by a higher percentage of households

(38.9%) and the financial situation is the reason least pointed out by the group, representing only 27.8% of households.

This shows that there is a relationship between the reason for building in downtown Ricatla and the perception of the harmful effects of flooding.

4.2.4 Causes of flooding

The main causes of flooding cited include heavy rainfall (46%), 2% and 3% of households say the cause is construction in low-lying areas and climate change, respectively. These causes reflect the combination of natural and anthropogenic factors that increase the risk of flooding in the area.

The high frequency of intense rainfall and soil compaction are natural factors that are difficult to control, but the accumulation of garbage and the removal of vegetation are issues that can be mitigated with proper environmental management practices and community awareness.

Figure34 : Graph of the causes of flooding as a percentage of households

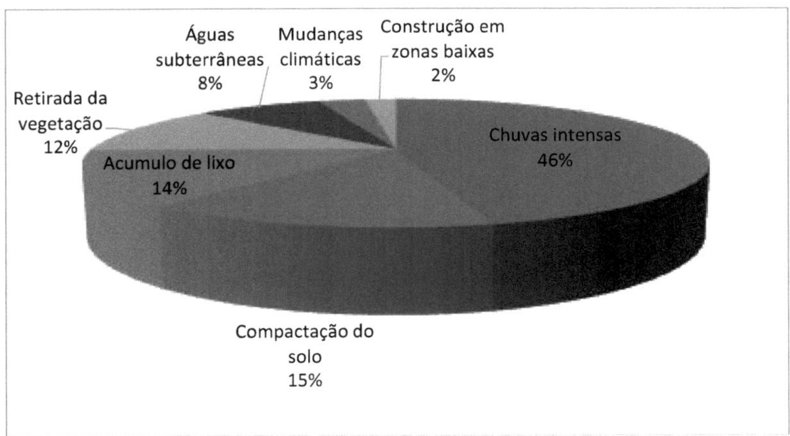

Source: Author, 2024.

4.3 Evaluation of acceptability

In order to assess the acceptability of the risk, we used parameters that could help measure the level of acceptance, such as: flood warning, where the interviewees recognize that there is a warning system, even so the attitudes taken are diverse in term of 30% of households remaining on the site, around 40% of households have already left the site and returned even though they know they have built in a risk zone and justify it by the lack of conditions 61% of households and these are the ones that have mostly been allocated spaces and sold.

If the risk is tolerable or unrestricted, this means that there is no need to adopt mitigating measures, unless you can reduce the risk further with little cost or effort. When it is tolerable with restrictions, it means that mitigating actions are recommended to reduce the risk and make it acceptable. Finally, when the risk is intolerable, it means that the current conditions must be changed until, at least, the risk is reduced to the level of tolerable with restrictions (Rocha, 2005).

The risk management process is divided into two main phases with their own characteristics: Risk assessment, which briefly leads to estimating the value of the risk and determining its acceptability, and is thus a decision support phase, where vulnerabilities are identified and the system's performance is analyzed in the face of hypothetical negative scenarios; and risk control, which seeks to reduce the risk (before the disaster) and control its consequences (after the disaster), by carrying out preventive actions, or maintaining risk and protection levels, with the aim of increasing security, both from a social and economic point of view, including, in particular, civil protection actions. It is in this last phase that the decisions taken on the basis of the risk assessment are implemented.

4.4 Acceptability indicators

To assess the acceptability of the risk of flooding in downtown Ricatla, the following indicators were established based on the responses of the interviewees and the observation of local patterns:

4.4.1 Flood alert

All the interviewees confirmed the existence of a flood warning system. However, the effectiveness of this alert was considered to be variable, depending on the speed of communication and the ability of families to react in time. Most residents recognize the importance of the alert, but many face difficulties in leaving their homes in good time.

4.4.2 History of abandonment

The survey results show that 40% of respondents have left the area at least once due to flooding. In contrast, 22% said they had never left their homes, even after the warnings. The remaining group, 38%, indicated that they always leave downtown Ricatla when they receive a warning of imminent risk. These data indicate a diversity of responses to danger, with factors such as risk perception and acceptability, family structure and financial conditions influencing evacuation decisions as shown in the graph in figure 35.

Figure35 : Graph of dropout history as a percentage of households

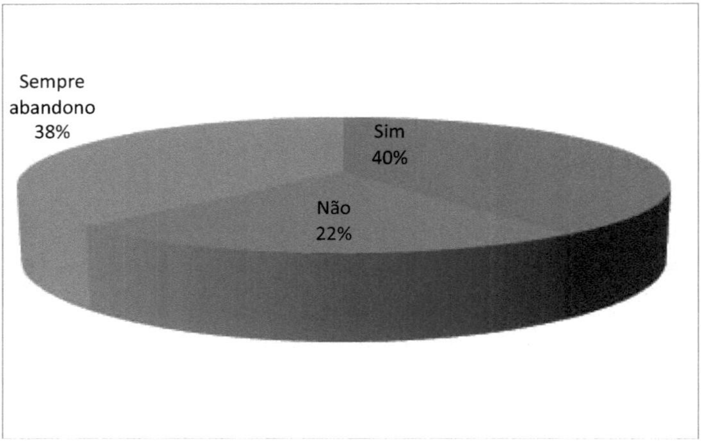

Source: Author, 2024.

4.4.3 Reason for return

The graph in figure 36 on the reason why households returned shows different reasons for returning to downtown Ricatla after the floods ended. The majority (61%) reported that they return due to a lack of financial conditions to settle elsewhere, highlighting the socio-economic vulnerability of the residents, while 20% mentioned that they returned for reasons linked to customs and family ties, reinforcing the importance of cultural identity and a sense of belonging.

For 17% of households, the return was motivated by the perception that the risk had decreased or was minimal, which demonstrates a subjective assessment of the seriousness of the situation. The rest returned for reasons related to work and studies, highlighting the lack of alternatives for these activities outside the risk area.

Figure36 : Graph of the reason for returning to the neighborhood as a percentage of households

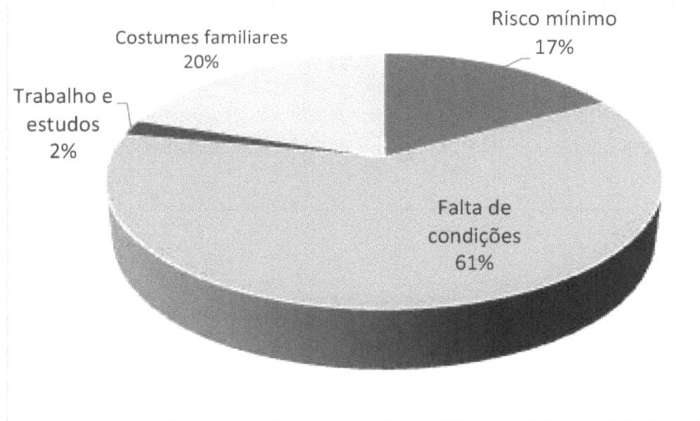

Source: Author, 2024.

4.4.4 Relationship between the reason for returning to the flood zone and the reason for building in the neighborhood

The table below shows that among the households that indicated minimal risk, lack of conditions and work and studies as the reason for returning to downtown Ricatla, a higher percentage in each of the groups indicated lack of financial conditions as the reason for building in downtown Ricatla; while those who pointed to family customs as the reason for returning to the flood zone affirmed, in equal percentage, the lack of alternatives and the fact of being native as the reason for building in downtown Ricatla (40%) and the financial situation was pointed out by only 20% of them.

Table 10: Relationship between the reason for returning to the flood zone and the reason for building downtown

	Reasons for returning to the flood zone			
	Minimal risk	Lack of conditions	Work and studies	Family customs

		N	%	N	%	N	%	N	%
Reason for buildin g in this neighb orhood	Financial situation	14	50.0 %	25	42.4 %	4	50.0 %	1	20.0 %
	Being native	5	17.9 %	16	27.1 %	2	25.0 %	2	40.0 %
	Lack of alternative s	9	32.1 %	18	30.5 %	2	25.0 %	2	40.0 %
Total		28	100.0 %	59	100.0 %	8	100.0 %	5	100.0 %

Source: Author, 2024.

4.4.5 Association between acceptability or perception and the reasons for building downtown

Table 11, on the relationship between acceptability or perception with the reasons for building downtown, both those who said yes and those who said no, most indicate the financial situation. In this case, there is no association between the households who answered yes and those who answered negatively.

The non-parametric chi-squared statistical test with a significance level of 5% resulted in a p-value of 0.643, higher than 0.05, indicating that there is not enough evidence to reject the null hypothesis. In other words, the reason for building in the neighborhood is independent of residents' knowledge or perception of the risk of flooding. Both those who consider the area a risk zone and those who do not have this perception build their houses mainly due to their financial situation or lack of alternatives.

It also shows that the lack of financial resources is the dominant factor both among those who recognize the risks of flooding and those who don't perceive them. Among those interviewed who consider the neighborhood a risk zone, 42.5% mentioned the financial situation as the main reason for building there, while 31.0% pointed to the lack of alternatives as the reason for living in the area. Locals represent 26.4% of this group, highlighting a minority who, although aware of the risks, remain in the neighborhood due to cultural or family ties.

The situation of the neighborhood's natives is also relevant, since many of them, despite recognizing the risks, remain due to historical or social ties. This suggests that any intervention to relocate this population or mitigate the risks must take these cultural factors into account, as well as considering economically viable housing solutions, as shown in table 11.

Table10 : Association between acceptability or perception and reasons for building downtown

			Reason for building in this neighborhood			Total
			Poor financial situation	They are native	Lack of alternatives	
You consider this place a risk zone	Yes	N	37	23	27	87
		%	42.5%	26.4%	31.0%	100.0%
	No	N	7	2	4	13
		%	53.9%	15.4%	30.8%	100.0%

P-value = 0.643 > 0.05

Source: Author, 2024

Table 12, on the association between attitude to flooding, shows that among those who say they leave the area, the majority are those who cannot afford it (62.3%); among those who say they remain in the area, 57.1% of households say they cannot afford it; of the households that put mechanisms in place, 50% say they return when the risk is minimal. Thus, there is no association between the reason for returning or staying in risk zones and the attitude towards warning, since the attitude towards warning is not determined by the reason for returning or staying in risk zones.

Table11 : Association between attitude towards floods

			Reason for returning to or staying in risk zones				Total
			Minimal risk	Lack of conditions to leave the neighborhood	Lack of housing alternatives	Family customs	
Attitude towards the alert	Abandoning the site	N	18	38	4	1	61
		%	29.5%	62.3%	6.6%	1.6%	100.0%
	I remain	N	8	20	3	4	35
		%	22.9%	57.1%	8.6%	11.4%	100.0%
	I create precautionary mechanisms	N	2	1	1	0	4
		%	50.0%	25.0%	25.0%	0.0%	100.0%

P-value = 0.233 > 0.05

Source: Author, 2024.

The chi-squared test with a p-value of 0.233, greater than 0.05, indicates that there is not enough evidence to reject the null hypothesis. Thus, the attitude of residents to the flood warning is not significantly associated with the reasons that lead them to return or remain in the area.

4.5 Relationship between risk perception and acceptability

The predominance of young people and men in Ricatla may be linked to the greater demand for housing space in areas of urban expansion, where the purchase of land is less expensive for the population than in more consolidated areas. The urbanization policy promoted by the Mozambican government has encouraged the expansion of urban areas, often resulting in disorderly construction and the occupation of risk zones. The pressure to accommodate a growing population often leads to inappropriate occupation of areas prone to disasters such as flooding, even under the supervision of local authorities.

In addition, the tendency to build in risk areas reflects the economic hardship faced by the population, who, despite being aware of the risks associated with flooding, are forced to choose sites with lower acquisition and maintenance costs, prioritizing immediate financial viability over long-term safety.

4.5.1 Socio-economic and housing conditions

Most of the families in lower Ricatla live in block houses, some in wood and zinc houses, reed houses and very few in matope houses. Practically everyone has access to drinking water, usually through boreholes and wells, with only a small percentage benefiting from water supply systems. The most widely used source of energy is coal, and everyone has limited access to health and education services.

The number of houses in the study area has grown without complying with any land-use planning instrument. The disorderly occupation, which is illegal, does not comply with environmental conservation and management standards, nor does it respect the normal plot shape (rectangular) adopted in many residential areas.

There was also a great deal of social inequality among the residents, caused by the irregular occupations as well as the size of the plots and the space occupied (in environmentally sensitive areas) as well as the material used to build the houses.

In addition, the quality of the buildings in downtown Ricatla is alarming, as most of them are unfinished and the walls are crumbling under the constant assault of water due to recurrent flooding. This reality is even more disheartening for the families who live in precarious housing made of wood, zinc or reeds, housing between 5 and 6 people in spaces consisting of just one bedroom and one living room.

As reported by some interviewees, access to water is often intermittent, and its quality varies depending on the maintenance of the reservoirs. In addition, the prices charged raise questions, since there doesn't seem to be a direct relationship between the time of supply and the quantity consumed, raising concerns about fairness in access to this vital resource.

This precarious condition of the housing, coupled with the instability of the water supply and the lack of clarity in the rectification criteria, highlights the urgency of interventions and policies aimed at improving housing conditions and guaranteeing equitable access to basic services, thus contributing to raising the standard of living of these vulnerable communities.

4.5.2 Length of residence in the lower Ricatla area

Around 50% of those interviewed have lived in the area for less than 5 years, 29% for more than 10 years, while the remaining 21% live in the neighborhood for between 5 and 10 years. The latter correspond to locals who have somehow returned to their homes even after being resettled at some point. Some of the locals confirm that under no circumstances will they leave the area, even though they have been through so many difficult times due to the disasters caused by the floods in the area under study.

Analysis of the ways in which plots were acquired in Ricatla reveals that 70% of residents acquired their plots through outright purchase. Other residents obtained their plots through inheritance or through bona fide donations from family members and acquaintances. Land prices vary between 20,000 and 150,000 meticais, depending on location, with land in elevated areas less prone to flooding being the most expensive.

This variation in prices suggests that, despite knowledge of the risks associated with flooding, many residents are attracted by the affordability of land in higher risk areas. The offer of land at lower prices in vulnerable areas can be a determining factor in choosing where to build, especially for those facing financial constraints.

4.5.3 Risk perception

With regard to the risk perception parameter, when questioning the interviewees about the reasons for building in lower Ricatla, approximately 44% of the participants revealed during the interviews that the choice was linked to their financial situation. Another 31% claimed a lack of alternative options, while the remaining 25% said they were natives of lower Ricatla. However, when it came to being victims of flooding, around 82% admitted to having already faced such a situation, while 18% were unaware of the risks of building and living in that area.

Similarly, 26% of the interviewees shared the perception that the risk of flooding in Ricatla is low, 30% considered it medium and 44% classified it as high. Regarding the level of the water during floods, 46% reported that the water reaches up to their knees, 40% said that it reaches their waist, and 14% mentioned that it reaches ground level. As for the causes pointed out by the interviewees, 46% believe that flooding is due to heavy rainfall, while 15%, 14%, 12% and 2% mentioned soil compaction, accumulation of garbage, removal of vegetation and construction in low-lying areas, respectively

Asked when floods are considered a risk. 62% of our households answered that it is when they threaten the property of any community, giving them a greater perception than the others who consider irreversible effects, from the point of view of when they have the power to kill or when they flood any place. Floods are therefore not always seen as a risk. Considering the potential catastrophe and irreversible effects is synonymous with a lower perception of the risk of flooding.

Cross-referencing this variable with the age and level of education variables, it can be seen that individuals aged between 18 and 24 with up to secondary education consider that there is a risk of flooding, but they can't do anything about it because they depend exclusively on their parents and guardians:

I'd like to live in a place where we don't have to climb on tables to sleep and whenever it rains we have to go out and sleep at our cousins'... we always have to buy new notebooks because if we're careless, they fall into the water and get wet.... (Interviewee 80)

One of the teenagers *in* the interview said *that it's always like this here, we've got used to it, the water doesn't even dry up completely, because by the time it starts to dry up it's November, it rains again until February, sometimes it's good for us because we can fish... (Interviewee 85).*

From these testimonies, it is clear that in Baixa de Ricatla there is a clear tendency for some households to accept the risk of flooding. This is related to the local population's recognition of the impossibility of totally eliminating the risk, as it recurs successively, and they also admit that it is a business opportunity (fishing) to live in this place, an idea shared by people over the age of 50.

The voluntary acceptance defended by Slovic (1978) as the ability to voluntarily accept risk because it is closely related to the perceived benefits is evident, and these benefits listed by the residents of downtown Ricatla are:

- Agricultural activities;
- Fishing activities;
- The proximity of the EN1 national road;
- Proximity to health and educational facilities;
- Proximity to the Zimpeto wholesale market;
- Having a house.

According to the results of the non-parametric chi-squared statistical test, the acceptability or perception of the risk of flooding by Ricatla residents is associated with their financial situation and lack of alternatives. The households interviewed categorically stated that they are aware of the risk they run in downtown Ricatla, but they become hostage to their own financial situation and other activities carried out near the site and prefer to stay.

Conclusion

The study on the acceptability of the risk of flooding in Lower Ricatla revealed that flooding is the result of a combination of natural factors, such as the terrain and the high water table, and anthropogenic factors, derived from poor land use and occupation, which leads to the silting up of drainage channels and an increase in surface runoff.

The sociodemographic analysis revealed that the majority of the residents of downtown Ricatla are young and male, mostly on low incomes with a level of education ranging from basic to higher.

Housing conditions in downtown Ricatla are precarious and unequal. The disorderly construction, with the predominance of materials such as wood, zinc and reeds, reflects the lack of proper urban planning and the occupation of environmentally sensitive areas. Most homes are incomplete and vulnerable to flooding, with the majority of structures being inadequate to cope with the impact of water, resulting in a compromised quality of life for residents.

Access to basic services such as drinking water and health care is limited, with many families relying on wells and boreholes for water and facing restrictions on access to medical care and education. There is social inequality among residents, evidenced by the variation in housing conditions and building materials. Negative impacts include the invasion of contaminated water (from latrines) into homes, the alteration of the natural course of river water, the reduction of the soil's capacity to filter surface water and the removal of the soil's fertile layer.

The length of residence reveals that approximately 50% of those interviewed have lived in the area for less than 5 years, suggesting that new residents may not be fully aware of the risks associated with flooding. Despite warnings and recommendations from the authorities, the lack of affordable housing alternatives and the high cost of land in safer areas mean that many residents opt for higher risk areas. The persistence of native residents returning to the area even after resettlement indicates a strong emotional and economic connection to the area, despite the challenges.

Analysis of how the plots were acquired shows that the majority of plots were purchased outright, with prices varying according to the risk of flooding. This indicates that although residents are aware of the risks, financial conditions and the lack of safer options lead many to choose cheaper plots in vulnerable areas. The variation in prices reflects the economic pressure on families, who are forced to prioritize affordability over long-term security.

The perception of risk among residents is varied, with approximately 82% admitting to having already faced floods, and 44% classifying the risk as high. Acceptance of risk and adaptation to the flooding environment are evident, with some residents considering flooding to be an inevitable part of their lives and even an opportunity for activities such as fishing. This voluntary acceptance of risk is related to the perceived benefits and the impossibility of completely eliminating the risk. Residents value aspects such as proximity to services and economic opportunities, despite the associated risks.

The acceptability of risk is also driven by the precarious economic situation of many residents. The lack of financial resources limits their housing options, forcing them to occupy risky areas. The informal economy, which prevails in the region, means that many residents depend on nearby locations for daily sustenance, making relocation to safer areas even more difficult. The affordability of land in risk zones is a temptation for those who would not otherwise be able to afford property in safer areas.

The study shows that there is no relationship between the association of acceptability or perception with the reasons for building in the downtown area: 42.5% of households say they are aware that they are living in a flood risk area and 53.9% of households say they are not. Among both those who said yes and those who said no, the majority pointed to the financial situation. This means that if they had better conditions, they wouldn't be living in downtown Ricatla.

The chi-squared statistical test carried out in this study showed that there is no statistically significant relationship between the reasons for building in the neighborhood and the perception of risk. This means that, regardless of whether residents are aware of the risk of flooding, the decision to stay or build in the area is strongly linked to their financial situation or the lack of housing alternatives.

The results confirm that, regardless of risk perception, socio-economic vulnerability is the main determinant in the choice to build in downtown Ricatla. The lack of financial conditions and the scarcity of housing alternatives limit residents' ability to choose, forcing them to remain in an area at risk, even when they are fully aware of the dangers associated with flooding

The work shows that economic vulnerability is closely linked to exposure to the risk of flooding, since these families don't have enough resources to choose locations with better infrastructure or away from risk areas.

Suggestions

   a) That the SDPI/Government of Marracuene removes families who have built in high-risk areas;

   ❖ That a safe area be identified, a detailed plan drawn up and infrastructural conditions created to provide various services in order to make it a pole of attraction for the families to be resettled;

   ❖ The vacant area should be turned into a public utility area. To this end, the process of transfer to the new area should be followed by demolition of the heritage built by the former residents.

   ❖ The implementation of mechanisms to inform and mobilize the community, based on the socio-environmental problems in the neighbourhood and/or study area, as a way of achieving collaboration with public initiatives and improving the quality of life;

❖ Establish effective communication channels between the community and the authorities, such as regular meetings and digital platforms;

❖ Implement a continuous monitoring system to evaluate the effectiveness of resettlement and development measures;

❖ Invest in the construction of drainage systems and containment barriers to reduce the impact of flooding;

❖ Implement reforestation projects and recover degraded areas to improve water absorption and reduce the risk of flooding;

❖ Encourage the use of durable building materials and construction techniques adapted to local conditions to increase the resilience of buildings.

Final Bibliography

1. ABREU; N. & ZANELLA; M.. *Perception of Flood Risks: Case Study in the Guabiraba Neighborhood, Maranguape - Ceará*. Ceará, 2015

2. ALMEIDA, M.A. *Acceptability of Risks in Industry, Brazil*, S/data.

3. ALMEIDA; A. *The Concept of Socially Acceptable Risk as a Critical Component of Risk Management Applied to Water Resources. Lisbon, S/data*;

4. ALÍPIO; Jaime & VALE Manuel. General Psychology. Anilda Ibrahimo Khan, Maputo S/data;

5. BORGESN. S. Urban Flood Risk Management. Coimbra, 2013.

6. BRAGA, J. O. Alagamentos e inundações em áreas urbanas: estudo de caso nacidadedesantamaria-(2014)DF.Universidade de Brasília; Brasília, 2016.

7. ALVES, Ramon Texeira Marques. *Construction of a Flood Map in an Urban Area*. Ouro Branco, 2019.

8. AMARAL, R; RIBEIRO, R. R. *Floods and inundations*. In: TOMINAGA, L.K; SANTORO, J; AMARAL, R. (eds.). *Natural disasters: knowing to prevent*. São Paulo: Geological Institute, 2009.

9. ARAÚJO & SANTOS. Perception of Flood Risk in Rondonópolis - MT. Brazil, 2020.

10. ASANTE, K.; et al. Socioeconomic implications of flood events: A global perspective. Natural Hazards, v. 42, n. 2, p. 345-360, 2007.

11. ASCROFT, J. História dos Riscos: Da Antiguidade à Modernidade. Lisbon: Editora Risco, 2001.

12. BARBOSA, F. *Measures for the protection and control of urban flooding in the Mamanguape River Basin/Pb*. Dissertation Federal University of Paraíba Technology Center, 2006.

13. BEILFUSS, R.; DOS SANTOS, D. Patterns of vegetation change in the Zambezi Delta, Mozambique. Working Paper No. 3, Zambezi Basin Wetlands Conservation Program, 2001.

14. BERNSTEIN, P. L. Against the Gods: The Remarkable History of Risk. Rio de Janeiro: Campus Publishing, 1997.

15. BEZERRA, A. C.. Risk Acceptability Criteria and Security Management. Rio de Janeiro: Editora Técnica, 2011.

16. BONIFACIO, A. Positive impacts of flooding: A contextual analysis. Environmental Perspectives, v. 30, n. 4, p. 567-582, 2022.

17. BORGES N. S. *Urban Flood Risk Management*. Coimbra, 2013.

18. BRAGA, J. O. *Alagamentos e inundações em áreas urbanas: estudo de caso na cidade de santa maria* - (2014) DF. University of Brasília; Brasília, 2016.

19. CAMPOS, S. J. A. M. Mapping areas subject to flooding for territorial planning and management: susceptibility, hazard and risk maps. Brazilian Journal of Environmental Engineering Geology. Page 67-81. 2015.

20. CARVALHO L. and SARAIVA M. G.. *Rivers and Cities: opportunities for urban sustainability*. Lisbon, 2009;

21. COCURULLO, A. *"Corporate Risk Management: Risks Aligned with Some Management Tools"*; 29th Edition, São Paulo, 2003;

22. COUNCIL OF MINISTERS. Master plan for disaster risk reduction 2017-2030. Approved by the 36th ordinary session of the Council of Ministers, 17 Oct. 2017. Available at: https://www.ingd.gov.mz/wp-content/. Accessed on: Aug. 5, 2024.

23. CRISTO, C. Natural Disaster Prevention: Strategies and Policies. São Paulo: University Press, 2004.

24. D'ARAÚJO; Rita. Seveso Directive Risk Acceptability Criteria for Portugal. Portugal, 2013

25. DE MOEL, H.; AERTS, J. C. J. H. Assessing flood damage and health risks: An integrated approach. Environmental Management, v. 35, n. 4, p. 567-582, 2009.

26. DOMÈNECH, C.; SUPRANAMIAM, R.; SAURI, D. *Risk Information and Adaptive Behavior*. Barcelona: Planeta Publishing House, 2010.

27. EL RAEY, M.; BEORO, A. Prescribed flooding and restoration potential in the Zambezi Delta, Mozambique. Working Paper No. 3, Zambezi Basin Wetland Conservation Program, 2003.

28. EL-RAEY, M.; BEORO, A. Assessing flood impacts: A case study in the Limpopo and Incomati basins. Water Resources Management, v. 25, n. 4, p. 567-582, 2003.

29. EMA. Flooding and its effects: An analysis of tangible and intangible impacts. Journal of Hydrological Sciences, v. 15, p. 35-48, 2002.

30. FELIZARDO, L. *Floods and inundation: natural phenomena and impacts on urban areas*. University Press, 2016.

31. FERNADEZ, Paulo. Flood risk assessment in urban areas with the integration of lidar data and large-scale cartography. Portugal, 2015;

32. GIULIO, G. et all. *Risk perception: a field of interest for the environment, health and sustainability interface*. São Paulo, 2015.

33. GOTHAM, K. F. Flooding, Risk, and Inequality: Urban Governance in the 21st Century. Cambridge: Polity Press, 2018.

34. GRACIOSA & MENDIONDO. *Flood Risk Management in the Context of Brazilian Urban Basins*. São Paulo, 2018.

35. Holz, L. K. "Urban Flooding: A Review of Causes and Control Measures." Journal of Urban Planning and Development, vol. 136, no. 4, 2010, pp. 227-238. DOI: 10.1061/(ASCE)0733-9488(2010)136:4(227)0733-9488(2010)136:4(227)).

36. INJAGE; L. *Social Perception of Flood Risk. The Case of Muriwa Village, Mopeia District (Lower Zambezi)*. Maputo, 2013;

37. NATIONAL INSTITUTE FOR DISASTER MANAGEMENT (INGD). Report on disaster management and the impacts of floods in Mozambique. 2014. Available at: https://www.ingd.gov.mz. Accessed on: July 5, 2024.

38. JINCHI, Z. *Economic consequences of flooding: A comparative analysis of urban and rural areas. Environmental Economics,* v. 30, n. 3, p. 123-138, 2005.

39. JINCHI, Z. *Patterns of hydrological change in the Zambezi Delta, Mozambique.* Working Paper No. 3, Zambezi Basin Streams and Wetlands Conservation Program, 2005.

40. KELLENS, W., *et al. "An analysis of public perception of flood risk on the Belgian coast".* Belgian, 2011.

41. KHALEQUZZAMAN, M. Impacts of flooding: A study on the Zambezi River Basin. Journal of Environmental Studies, v. 10, n. 2, p. 45-62, 1994.

42. KLAIS, R. Perception and Sensation: Understanding the Environment. São Paulo: Editora Cognitiva, 2012.

43. KOBIANA, J., MENDONÇA, J., & MORENO, A. The impacts of flooding: an analysis of tangible and intangible effects. Journal of Environmental Studies, 10(2), 45-62. 2006.

44. KOBIYAMA, M. et al. Prevention of natural disasters - basic concepts. Ed. Curitiba - PR, Organic Trading, 2006.

45. LE BETONE, M. Risco e Sociedade: Uma Perspectiva Contemporânea. São Paulo: Societas Publishing House, 2012.

46. LECHOWSKA. E. *What determines flood risk perception? A review of flood risk perception factors and relationships between their basic elements.* Poland 2018.

47. MACHADO, C. E. *Inundações Urbanas: Processos, Causas e Consequências*. Porto Alegre: Editora Clima, 2010.

48. MAE (Ministry of State Administration). *Marracuene District Profile*, 2014

49. MANUELE. F. *Acceptable Risk*. São Paulo, 2010

50. MATLOMBE, L. F. *Participation of Vulnerable Communities in Flood Risk Management in Lower Limpopo - Mozambique*. Dissertation for a Master's Degree in Sustainable Urbanism and Spatial Planning, Universidade de Nova Lisboa, 2019.

51. MATOS, J; JARDILINO, P. Perception and Representation: Cognitive Processes. Porto Alegre: Editora Psicologia, 2016.

52. MESSNER, D; MEYER, J. Risk Perception: Variability and Social Influences. Berlin: Risk Analysis Publishing House, 2005.

53. MESSNER, F. & MEYER, V. *Flood damage, vulnerability and risk perception: challenges for flood damage research*. UFZ, 2005.

54. MICOA (2005). Assessment of vulnerability to climate change and adaptation strategies. Available at http://fsg.afre.msu.edu/mozambique/caadp/Avaliacao_vulnerab_mud_ climat_estrateg_adapt .pdf. Accessed March 2024.

55. MICOA (2006). *Mozambique, Improvement of Informal Settlements, Analysis of the Situation Proposal for Intervention Strategies*. Maputo

56. MICOA. Intervention Strategy for Informal Settlements in Mozambique. Maputo, 2010.

57. MIRAMAR. Reports from the head of the Ricatla neighborhood about floods. 2024. Available at: https://www.youtube.com/watch?v=2jx6XgN4dX4. Accessed on: 5 March 2024.

58. MONTE, B E O, et all. *Hydrological and Hydraulic Modeling to Floodplain Mapping*. Porto alegre, brazil, 2016;

59. MOTA, R.. "Urban Hydrology and Water Management-Present and Future Challenges." Water Science and Technology, vol. 47, no. 3, 2003, pp. 9-17. DOI: 10.2166/wst.2003.0077

60. MURWIRA, A. Health implications of post-flood water contamination: A case study in the Limpopo River Basin. Water and Health, v. 20, n. 3, p. 123-138, 2006.

61. NDAPASSOA, A. M. Mozambique's experience in disaster management and socio-environmental post-disaster recovery (2019-2023). Veredas do Direito, Belo Horizonte, v. 20, e202565, 2023. Available at: http://www.domhelder.edu.br/revista/index.php/veredas/article/view/2 565. Accessed on: July 15, 2024.

62. OLIVEIRA. F. Percepção de Riscos Ambientais e Mudanças Climáticas no Varjão Distrito Federal. Brasília 2012.

63. Land Use and Occupation Master Plan (PDUL). Maputo City Hall, 2020.

64. POSSE, J. S. *Educação Ambiental e Cidadania: Práticas e Desafios na Preservação Ambiental*. Rio de Janeiro: Editora Ambiental, 2011.

65. RAMOS, C. *Perigos Naturais a causas Meteorológicas: o Caso e Inundações*. Center for Geographical Studies, Institute of Geography and Spatial Planning. University of Lisbon. 2013

66. Republic of Mozambique. Decree no. 60/2006 of December 26 - Urban Land Regulations. In: Boletim da República, 2006. Available at: https://www.impacto.co.mz/legislacao-ambiental/. Accessed on: 05 May. 2024.

67. Republic of Mozambique. Law No. 10/2020 of August 14 - Disaster Risk Reduction and Management Law. In: Boletim da República, 2020. Available at: https://www.impacto.co.mz/legislacao-ambiental/. Accessed on: 15 November 2023.

68. Republic of Mozambique. Law No. 15/2014 of June 20 - Establishes the Legal Regime for Disaster Management. In: Boletim da República, 2014. Available at: https://www.impacto.co.mz/legislacao-ambiental/. Accessed on: 05 February 2024.

69. Republic of Mozambique. Law no. 19/1997 of October 1 - Land Law. In: Boletim da República, 1997. Available at: https://gazettes.africa/archive/mz/1997/mz-government-gazette-series-i-supplement-no-3-dated-1997-10-07-no-40.pdf. Accessed on: December 30, 2023.

70. Republic of Mozambique. Law no. 19/2007 of July 18 - Land Planning Law. In: Boletim da República, 2007. Available at: https://www.impacto.co.mz/legislacao-ambiental/. Accessed on: November 15, 2023.

71. Republic of Mozambique. Law No. 20/1997 of October 1 - Environmental Law. In: Boletim da República, 1997. Available at: https://gazettes.africa/archive/mz/1997/mz-government-gazette-series-i-supplement-no-3-dated-1997-10-07-no-40.pdf. Accessed on: January 31, 2024.

72. ROCHA J.. *Flood Risk and its Management. A National Vision and a European Vision.* Portugal. 2005

73. SANTOS, Milton. *Metamorfoses do espaço habitado: fundamentos teóricos e metodológicos da geografia.* 6. ed. São Paulo: Editora da Universidade de São Paulo, 2007.

74. SALINAS & ESPINOSA, *Settlements near* watercourses, Brazil 2004.

75. SILVA, A. A. & POLERO, B. B. Urban Flooding: Causes and Consequences. Urban Environment, 10(2), 45-62. 2018

76. SILVA, J. Risco e Modernidade: Uma Análise Histórico-Sociológica. Porto Alegre: Editora Moderna, 2017.

77. SILVA, K. C; POLERO, C. *Socio-environmental perceptions of floods: reflections on risk*. INTERthesis International Interdisciplinary Journal, 2015.

78. SILVA, M.L. *Percepção Ambiental e Participação Comunitária: Estudos e Aplicações*. São Paulo: Editora Terra, 2000.

79. SIMBINE, C. R. *Analysis of the correlation between population growth and loss of vegetation cover, case study: Marracuene District, 1997-2007-2017*. Scientific monograph, Maputo Pedagogical University - Faculty of Earth and Environmental Sciences, 2023.

80. SLOVIC, P. Perception of risk. Im Scieence, n° 236, 1987. 2010

81. SOUZA, A. P. Diagnosis *of Perception of Environmental Risks*. Dissertation presented at the University of Lavras for a Master's degree in 2015.

82. SOUZA, C. C. & ROMUALDO, D. D. Social Vulnerability in Risk Areas. In M. N. Santos (Org.), Natural and Urban Disasters (pp. 123-145). São Paulo: Editora XYZ.2020

83. SOUZA, F. *Environmental Education and Sustainability: Theories and Practices for the Preservation of Natural Resources*. São Paulo: Editora Verde, 2018.

84. SOUZA, L. B., & ZANELLA, M. E. Perception of Environmental Risks: Theories and Applications. Fortaleza: UFC Editions, 2009.

85. SPINK, P. K.; et al. *O Risco no Quotidiano*: Saúde, Trabalho e Meio Ambiente. São Paulo: Editora FGV, 2004

86. SUSAN, M.; MYERS, J.. *Discussing Risk: Ethical and Social Perspectives*. New York: Social Publishing, 2009.

87. TELEVISÃO DE MOÇAMBIQUE (TVM). Report on the risk of flooding in the Ricatla neighborhood, Marracuene. 23 Jun. 2019.

88. TEODORO, V. L. I. et al. The watershed concept and the importance of morphometric characterization for understanding local environmental dynamics. Revista Uniara. Araraquara, n.20, p. 137-15, 2007.

89. TUCCI, C. E. Urban rainwater management (p. 12). Programa de Modernização do Setor Saneamento, Secretaria Nacional de Saneamento Ambiental, Ministério das Cidade 2005.

90. TUCCI, C. E., & BERTONI, J. C. Urban flooding in South America. Ed. dos Autores, 2003.

91. TUCCI, C.E.M. *Inundações Urbanas.* Porto Alegre: ABRH/RHAMA.1999.

92. UNDP, Bureau for Crisis Prevention and Recovery. Reducing disaster risk: achallenge for development a global report. New York, 2004

93. UNISDR. Flood Hazard and Risk Assessment. In: Words into Action Guidelines: National Disaster Risk Assessment Hazard Specific Risk Assessment. From World Bank estimates, 2017.

94. VESTENA, L et all. *Environmental Perception on the Causes of Flooding, Guarapuava/Pr: in Search of a Resilient City.* São Paulo, 2014.

95. WOLLMANN, C. A. *River dynamics and their geographical consequences.* Revista de Geografia, 20(2), 45-62, 2015.

96.

Appendices

MAPUTO PEDAGOGICAL UNIVERSITY
FACULTY OF EARTH AND ENVIRONMENTAL SCIENCES
Master's Degree in Environmental Risk Management

INTERVIEW SCRIPT

> **Theme:**
>
> **Acceptability of Flood Risk in Lower Ricatla - Marracuene District**

Introducing the interviewer

They me Judite Pinto Tene, a final-year student on the Master's course in Environmental Risk Management at the Faculty of Earth and Environmental Sciences at the Pedagogical University of Maputo.

This interview is part of the culminating work for the course mentioned above, the subject of which is **"Acceptability of the Risk of Floods in Lower Ricatla, Marracuene District"**. It should be noted that this is a series of questions for purely academic purposes, so all information provided during the interview will be kept strictly confidential.

Therefore, I would greatly appreciate your cooperation in participating and answering these questions, as they will be of great value for the materialization of this work and for obtaining a degree in environmental education. However, I would be grateful if you could take an active part by providing truthful information on the subject under discussion.

Thank you very much for your attention!

The main aim of the questionnaire is to understand how the Marracuene District Planning and Infrastructure Service manages the issue of informal occupation and its impacts in order to guarantee the quality of life of its residents. The answers given will be for academic purposes only and the author is responsible for the confidentiality of the information provided.

1. It is known that in recent years the district of Marracuene has seen a sharp increase in the use and occupation of space/soil, with the supposed result being an accelerated process of urbanization.
    a) What are the parameters for access to land?
    b) What criteria are used to allocate spaces?
2. What do you do when you receive a request for space for various purposes?
3. Man, in his interaction with space, must promote sustainable development, which involves a balanced coexistence between man and the environment.
    c) How do you see the process of occupying space in Baixa do Ricatla?

Thank you very much for your attention!

1. How many people live in the neighborhood? ___________

2. What criteria have they taken into account when allocating a space/land to this person?

3. For what purposes do residents look for space/land in this neighborhood?

4. Is there a health center in the neighborhood? Yes____ No____

5. Is there a school in the neighborhood? Yes____ No____

    a) How many teach pre-primary level? _________

    b) How many teach at primary level? ___________

    c) How many teach at primary level? _________

    d) How many teach at secondary level? _________

6. What basic services does the neighborhood benefit from

7. What are the main problems affecting Baixa do Ricatla during the rainy season?

8. What are the main causes of this problem?

9. Have any homes or families been affected by this problem and had to leave their homes?

Thank you very much for your attention!

Name: ___________________________________   Neighborhood:

___________________ Block: __________

I. DATA

1. Sex

a) F |___| b) M |___|

2. interview location

a) Block   b) Neighborhood |________________|

3. Age range

a) 15-24 |___| b) 25-34 |___| c)35-444 |___| d) 45-54 |___| e) 55 + |___|

4. Marital status

a) Married |___| b) Single |___| c) Marital |___| d) Divorced |___| e) Widowed |___|

5. level of education

a) No schooling |___| b) Literacy studies |__|

c) Complete primary studies |__| d) Secondary studies |__|

e) High school |__| f) University |__| g). Other |__| Which? ___________________

6. What type of housing do you have?

a). Modern |___| b) Conventional |___| c) Rudimental |___|

d) Other |__| Which? ________________

7. How long have you lived in this neighborhood?

a) 1-2 |___| b) 3-5 |___| c)6-10 |___| d) 11-20 |___| e) +21 |___|

8.How was the plot acquired?

a) District Government |___| b) Purchase |___| c) Inheritance |___| d) Good faith |___|

e) Other |___| Which? _____________

9. Type of housing

a) Blocks |__| b) Wood and Zinc |__| c) Reed |__| d) Brick |__| e) Matope |__| d) Other
|__| Which ones?_____________

10. What are the sources of energy for domestic use?

a) Firewood |__| b) Coal |__| c) Oil |__| d) Electricity |__| e) Candle |__| f) Gas |__| g)
Other |__|. Which ones? _________________

11. What are the sources of drinking water?

a) Fountain |__| b) Borehole |__| c) Well |__| d) Other |___|. Which
ones?_________________

12. Do you have access to health services?

a) Yes |__| b) No |___| c) They go to other neighborhoods |___|. Which
one?_______________

13. Do you have access to education services?

a) EP1 |___| b) EP2 |___| c) ESG1 |___| d) ESG2 |___| e) ETP |___| f) Centro Infantil
|___| g) Superior |___| h) They go to other neighborhoods |___| i) Others |___|. Which?
—

14. Do you have access to security services?

a) Yes |___| b) No |___| c) They go to other neighborhoods |___|. Which one? ______

15. What impacts on the environment have been observed as a result of the use and
occupation of space in this area? ________________________________

16. What motivated you to build your house in this location? ________________ -

17. Has your home ever been affected by flooding? a) Yes |__| b) No |__|

18. What are the main causes of these events?

19. Do you consider this place to be at risk of flooding? a) Yes |__| b) No |__|

20. Who do you think is responsible for the risk of flooding in the neighborhood where you live? ________________________________

21. Do you think floods can cause damage? a) Yes |__| b) No |__|

22. When do floods occur each year?

a) Whenever it rains |__| b) Only during the rainy season |___| c) All year round |___|.

23. How do you rate the level of flood risk in this location?

a) High |__| b) Medium |__| c) Low |__| d) Does not exist |___|.

24. What is the water level when flooding occurs?

a) Reaches to the knees |__| b) Waist |__| c) Floor level |___|.

25. What is your level of knowledge about flood risks?

a) Not informed |__| b) Somewhat informed |__| c) Informed |__|.

26. Do you think the risk of flooding could have an effect in the future? a) Yes |__| b) No |__|

27. Who do you blame for the flooding in your neighborhood?

a) Intense rainfall |___| b) Soil compaction |___| c) Removal of vegetation |___|

d) Accumulation of waste |___| e) Construction in low-lying areas |___| f) Climate change |___| g) More shallow groundwater |___|.

28. Are you aware that you live in a flood risk area? a) Yes |__| b) No |__|.

29. When do you consider floods to be a risk?

a) When it floods any place |___| b) When it threatens the property of any community |___| c) When it has the power to kill |___| d) When it hits houses |___|.

30. How do you rate the possibility of water entering your house in the event of flooding?

a) Very high |___| b) High |___| c) Medium |___| d) Low |___|.

31. Is there a flood warning at this location? a) Yes |__| b) No |__|

32. Is there an institution that controls the risk of flooding in the area? a) Yes |__| b) No |__|.

33. What do you do when you receive a flood alert?

a) Leaving the scene |___| b) Remaining |___| c) Developing a precautionary mechanism |___|.

34. Have you ever left the area because of flooding?

a) Yes |___| b) No |___| c) I've always dropped out |___|.

35. Why did you return to the flood zones?

a) When the risk is minimal |___| b) Lack of conditions |___| c) Work and studies |___| d) Guarding family customs |___|.

36. What are the neighbourhood's initiatives to minimize the risk of flooding that affects you?

a) Opening drainage areas |___| b) Planting vegetation |___| c) Hiring institutions to remove water |___| d) No initiative |___| e) Other |__| Which?________________.

37. Under flood conditions, have you benefited from resettlement? a) Yes |__| b) No |__|.

38. If you were given land in a safe space, would you leave this place? a) Yes |__| b) No |__|

Thank you very much for your attention!

I want morebooks!

Buy your books fast and straightforward online - at one of world's fastest growing online book stores! Environmentally sound due to Print-on-Demand technologies.

Buy your books online at
www.morebooks.shop

Kaufen Sie Ihre Bücher schnell und unkompliziert online – auf einer der am schnellsten wachsenden Buchhandelsplattformen weltweit! Dank Print-On-Demand umwelt- und ressourcenschonend produziert.

Bücher schneller online kaufen
www.morebooks.shop

Printed by Books on Demand GmbH, Norderstedt / Germany